Space Law and Policy

Series Editor

Maria A. Pozza, Gravity Lawyers, Christchurch, New Zealand

Space activities across the Asia-Pacific, specifically in Australia and New Zealand, are increasing and diversifying. Over the last decade, countries in this region have seen significant growth within their national space industries and economies which in turn benefit their wider societies due to the downstream application and use of, space-related technologies, products, and services. This phenomena presents not only a great opportunity for these countries, but also represents a situation necessitating considered and fit-for-purpose space law and policy. Robust yet flexible regulatory regimes are needed if countries in the Asia-Pacific, such as Australia and New Zealand, are to benefit from the strategic and commercial advantages associated with a presence in outer space and functioning domestic space sector. These regulatory regimes must respond to, and address, numerous challenges ranging from adherence to international legal obligations to issues of geo-political space security and military concerns through to questions about how best to regulate and support commercial space endeavours. These myriad factors will have an impact upon how countries develop, implement, and refine their space laws and policies. This series offers a unique perspective from this region and intends to act as a source of knowledge and guidance for space actors across government, the private sector and academia in the pursuit and furtherance of their respective space activities and operations. The contributors to this series are all experts in their field who share a common interest in strengthening space governance via the promotion and development of space law and policy.

Hamid Jahankhani · Stefan Kendzierskyj ·
Sina Pournouri · Maria A. Pozza
Editors

Space Governance

Challenges, Threats and Countermeasures

Springer

Editors
Hamid Jahankhani
Northumbria University London
London, UK

Stefan Kendzierskyj
Cyfortis
Worcester Park, UK

Sina Pournouri
Sheffield Hallam University
Sheffield, UK

Maria A. Pozza
Gravity Lawyers
Hornby, Christchurch, New Zealand

ISSN 2662-9054 ISSN 2662-9062 (electronic)
Space Law and Policy
ISBN 978-3-031-62227-4 ISBN 978-3-031-62228-1 (eBook)
https://doi.org/10.1007/978-3-031-62228-1

This Springer imprint is published by the registered company Springer Nature Switzerland AG
The registered company address is: Gewerbestrasse 11, 6330 Cham, Switzerland

Contents

Space Governance Frameworks and the Role of AI and Quantum Computing

Stefan Kendzierskyj, Hamid Jahankhani, and Osama Akram Amin Metwally Hussien

Abstract Satellites have advanced from scientific wonders to vital components of our daily lives in modern society. These defined "artificial objects" provide worldwide communication, accurate navigation, Earth observation, national security and many more crucial services in our interconnected world. However, as our reliance on satellite technology grows, so do cybersecurity concerns. The growing dependence on satellite technology necessitates the development of a comprehensive governance policy or framework to effectively oversee satellite operations, management, and data transmission. The rapid increase in satellite launches by governments, commercial organisations, and international organisations has prompted concerns regarding space debris, orbital congestion, and potential security risks. The demand for satellite governance is driven by the diminishing dimensions and exponential proliferation of satellites within the Earth's orbit. Satellite structures are susceptible to the presence of vulnerabilities that can be exploited in their nodes and linkages. The importance of satellite ground systems in the context of cybersecurity is frequently disregarded in conversations about satellite security. This chapter aims to critically review the current state of the research in the field and examine the establishment of a global regulatory framework for satellites, given the current absence of such a framework. This chapter will also examine the impact of AI-based cyber-attacks and the role of quantum computing on satellites and the attempt to propose feasible mitigating strategies.

Keywords Satellites · Space · AI · Quantum · LEO · MEO · HEO · Cybersecurity · Debris · NASA · ITIL · COBIT

S. Kendzierskyj · H. Jahankhani (✉) · O. A. A. M. Hussien
Northumbria University London, London, UK
e-mail: Hamid.jahankhani@northumbria.ac.uk

H. Jahankhani et al. (eds.), *Space Governance*, Space Law and Policy,
https://doi.org/10.1007/978-3-031-62228-1_1

1 Introduction

Cybersecurity and satellite communications are interrelated and critical domains facing various challenges and threats in the contemporary world. Satellites play an important role in all aspects of modern life, orbiting the Earth and improving our ability to monitor the environment, link to the rest of the globe and protect national security. However, as our dependence continually grows on this technology, so do the associated risks, especially with cybersecurity, AI and other factors that make these artificial objects that orbit Earth attractive targets to malicious actors.

Satellite communications enable global connectivity, navigation, and remote sensing for various sectors and applications, such as military, civil, commercial, and scientific. However, they also depend on secure and resilient cyber capabilities for their functions, operations, and transmissions. Cybersecurity is the protection of information systems and networks from unauthorized access, use, modification, or destruction. It involves technical, operational, strategic, and legal aspects that aim to ensure the confidentiality, integrity, and availability of data and services. Satellites and ground stations can be a hotspot for these types of data and services; thus, they are also vulnerable to cyber-attacks that can disrupt or damage their communications and services. These attacks can be kinetic, virtual, or hybrid, and can target satellites or their ground stations by hacking, jamming, spoofing, or denial-of-service. Kinetic attacks involve physical damage or destruction of satellites or their components by missiles, lasers, or space debris. Virtual attacks involve cyber operations that target the software, data, or signals of satellites or their ground stations by hacking, jamming, spoofing, or denial of service (DoS). Hybrid attacks combine both kinetic and virtual elements to achieve a synergistic effect. Such attacks pose serious risks to the security, economy, and society of the affected states and require effective countermeasures and legal frameworks.

Various research papers have addressed these issues from different perspectives by focussing on:

- The risk assessment of cyber-attacks on satellite communications, using quantitative methods such as fault tree analysis or Bayesian networks to estimate the probability and impact of different attack scenarios.
- The resilience engineering of satellite communications, using qualitative methods such as system dynamics or resilience matrix to identify the vulnerabilities and capabilities of satellite systems and networks.
- The cyber deterrence of satellite communications, using game-theoretic models such as signalling games or extended deterrence games to analyse the strategic interactions and outcomes of different actors involved in cyber conflicts.
- The normative development of satellite communications, using legal analysis and policy recommendations to address the gaps and challenges in the existing international law regimes that apply to cyber-mediated ASAT events.

The research highlights the need for a comprehensive and integrated approach that considers the technical, operational, strategic, and legal aspects of cybersecurity and satellite communications. They also point out the limitations and difficulties of applying the existing legal frameworks to the emerging threats and challenges in this field. There are current gaps in emerging cybersecurity frameworks and international space laws and propose a global framework for cooperation and coordination among relevant stakeholders to ensure the security and sustainability of satellite systems and communications, is a viable necessity. It will require exploration of the intersections and gaps between the existing legal regimes and the nascent cybersecurity norms, and suggest some elements for a normative framework that can address the current and future challenges in this field and propose a methodology for counteracting the possible future cyber-attacks with the upcoming emergence of quantum computers and artificial intelligence.

1.1 Space History and Current Treaties

The first phase of space governance began in the late 1950s, when the launch of Sputnik 1, by the Soviet Union, triggered the space race between the two superpowers of the Cold War era. The United Nations (UN) became the main forum for discussing and negotiating the legal and political issues related to outer space, as well as for promoting international cooperation and peaceful utilisation of space. The UN established two main bodies to deal with space matters: the Committee on the Peaceful Uses of Outer Space (COPUOS) and its two subcommittees (the Scientific and Technical Subcommittee and the Legal Subcommittee), and the Office for Outer Space Affairs (OOSA), which serves as the secretariat for COPUOS and its subcommittees, as well as for other UN entities involved in space activities. The main outcome of this phase was the adoption of five international treaties that constitute the core of space law, shown in Table 1:

1.2 Non-Binding and Binding Agreements in Space Governance

There are two primary classifications of space governance:

- Non-binding agreements employed to convey voluntary, non-normative, and/or aspirational content that may prove difficult to achieve global consensus on.
- Binding or normative instruments encompassing treaties, norms, and national regulations.

These two categories greatly enhance each other and make substantial contributions to the current framework for composite space governance. Presently, there is a notable expansion in the range and diversity of activities and entities within the space

Table 1 Adoption of five international treaties that constitute the core of space law

International treaties	Conditions of the treaties
The Treaty on Principles Governing the Activities of States in the Exploration and Use of Outer Space, including the Moon and Other Celestial Bodies (the "Outer Space Treaty")—adopted on 19 December 1966 in General Assembly resolution 2222 (XXI), opened for signature on 27 January 1967, entered into force on 10 October 1967	The Outer Space Treaty is considered to be the cornerstone of space law, as it establishes the fundamental principles that govern all aspects of space activities, such as: 1. The freedom of exploration and use of outer space by all states without discrimination 2. The prohibition of placing weapons of mass destruction or establishing military bases in outer space or on celestial bodies 3. The non-appropriation of outer space or any part thereof by any means 4. The responsibility and liability of states for their national activities in outer space, whether carried out by governmental or non-governmental entities 5. The obligation to conduct all activities in outer space with due regard to the interests of other states and in accordance with international law 6. The obligation to cooperate with other states in the exploration and use of outer space for peaceful purposes 7. The obligation to render assistance to astronauts in distress and to return them to their state of registry 8. The obligation to notify and consult with other states on potentially harmful interference with their activities in outer space 9. The obligation to register objects launched into outer space with an appropriate national authority and with the UN Secretary-General
The Agreement on the Rescue of Astronauts, the Return of Astronauts and the Return of Objects Launched into Outer Space (the "Rescue Agreement")—adopted on 19 December 1967 in General Assembly resolution 2345 (XXII), opened for signature on 22 April 1968, entered into force on 3 December 1968	The Rescue Agreement elaborates on some provisions of the Outer Space Treaty regarding astronauts and space objects, such as: 1. The duty to take all possible steps to rescue and assist astronauts in distress on land or at sea, or in outer space or on a celestial body, regardless of their nationality 2. The duty to notify immediately the launching authority or state of registry, as well as the UN Secretary-General, about any personnel or objects encountered in outer space or on a celestial body 3. The duty to return promptly any personnel or objects launched into outer space to their state of registry upon request

(continued)

Table 1 (continued)

International treaties	Conditions of the treaties
The Convention on International Liability for Damage Caused by Space Objects (the "Liability Convention")—adopted on 29 November 1971 in General Assembly resolution 2777 (XXVI), opened for signature on 29 March 1972, entered into force on 1 September 1972	The Liability Convention specifies the rules and procedures for the settlement of claims for damage caused by space objects, such as: 1. The absolute liability of the launching state for damage caused by its space object on the surface of the Earth or to aircraft in flight 2. The fault-based liability of the launching state for damage caused by its space object to another space object in outer space 3. The joint and several liability of two or more states that jointly launch or procure the launching of a space object, or from whose territory or facility a space object is launched 4. The right of any state that suffers damage, or whose natural or juridical persons suffer damage, to present a claim for compensation to the launching state or states 5. The obligation of the states concerned to establish a claims commission if no settlement is reached through diplomatic negotiations within one year from the date of notification of the claim
The Convention on Registration of Objects Launched into Outer Space (the "Registration Convention")	This was adopted on 12 November 1974 in General Assembly resolution 3235 (XXIX), opened for signature on 14 January 1975, entered into force on 15 September 1976. The Registration Convention complements the Outer Space Treaty by requiring states to maintain national registers of objects launched into outer space and to furnish information on such objects to the UN Secretary-General, who maintains an international register. The information to be provided includes: 1. The name of the launching state or states 2. An appropriate designator of the space object or its registration number 3. The date and territory or location of launch 4. The basic orbital parameters, including nodal period, inclination, apogee, and perigee 5. The general function of the space object The Registration Convention aims to enhance transparency and accountability in space activities, as well as to facilitate the identification and tracking of space objects

(continued)

Table 1 (continued)

International treaties	Conditions of the treaties
The Agreement Governing the Activities of States on the Moon and Other Celestial Bodies (the "Moon Agreement")	Adopted on 5 December 1979 in General Assembly resolution 34/68, opened for signature on 18 December 1979, entered into force on 11 July 1984. The Moon Agreement applies the principles and provisions of the Outer Space Treaty to the Moon and other celestial bodies within the solar system, other than Earth, with some modifications and additions, such as: 1. The recognition of the Moon and its natural resources as the common heritage of mankind, which implies that their exploration and use should be carried out for the benefit and in the interests of all countries and humankind 2. The prohibition of any change in the environment of the Moon or any adverse effect on its environment resulting from human activities 3. The establishment of an international regime to govern the exploitation of the natural resources of the Moon when such exploitation becomes feasible 4. The obligation to inform and consult with other states parties about any activity or experiment planned or conducted on the Moon The Moon Agreement is the least ratified among these treaties, as it has been criticized for its vague and controversial provisions regarding the common heritage principle and the international regime for resource exploitation

domain. The establishment of robust legal and policy frameworks for these endeavours, necessitates the active involvement and cooperation of diverse stakeholder groups. Additionally, enhancing communication channels between the private sector and international forums responsible for space governance is imperative. However, the expansion is subject to significant uncertainty due to various commercial and technical factors.

Emerging activities require novel approaches to governance, which should align with the specific nature of the activities in question and avoid hindering the progress of innovation. In order to effectively meet international obligations, the establishment of legal certainty, common norms, and regulation is imperative. Achieving this equilibrium necessitates the implementation of regular modifications to regulatory rules and frameworks, rather than adopting a comprehensive approach towards novel applications. Additionally, it is imperative to establish channels for information sharing among various stakeholders, including the technological, economic, corporate, policy, and regulatory sectors. The concept in question pertains to governance rather than being tied to a particular structure or methodology. In light of the present safety-of-flight circumstances and the projected escalation of Space Traffic Management (STM) and Space Situational Awareness (SSA) requirements, along with the corresponding increase in safety risks due to the advent of the New Space era, it is apparent that a comprehensive strategy for space debris management is imperative (Oltrogge & Christensen, 2020).

1.3 Role of NATO and the UN in Space

The North Atlantic Treaty Organisation (NATO) is currently engaged in the development of a Strategic Space Situational Awareness System (3SAS) at its headquarters. This capability will enable the Alliance to enhance its understanding of the space environment, space events, and their impacts across various domains. While the primary focus of 3SAS lies in observing celestial bodies, the Alliance must allocate resources towards the development of assets that enable comprehensive monitoring of terrestrial activities.

In alignment with the aforementioned guiding principles, NATO's comprehensive space strategy will prioritise the subsequent fundamental functions. The primary objective will be to facilitate political-military discussions and the exchange of information on space developments related to deterrence and defence.

Its purpose is to improve the Alliance's understanding of the current situation, decision-making processes, preparedness, and overall management of its posture in various conflict scenarios and cover topics such as:

- Integrating spatial and space-related factors into the implementation of NATO's core obligations: collective defence, crisis management, and, where relevant, cooperative security.
- The aforementioned discussions may encompass various aspects such as threats, challenges, vulnerabilities, opportunities, and the development of ethical and legal standards in different domains.
- It is crucial to guarantee the effective delivery of space support and its impact on the operations, missions, and other endeavours of the Alliance.
- An important objective is to facilitate the enhancement of compatibility and interoperability among the space services, goods, and capabilities (NATO, 2022).

The Outer Space Treaty (OST) of 1967 emerged as a consequential outcome of the Space Race between the Soviet Union and the United States during the Cold War. This treaty established a comprehensive and enduring framework for international collaboration in space exploration, while emphatically prohibiting any assertions of national sovereignty over outer space. Nevertheless, as a consequence of the swift advancement of technology, the OST has become outdated. The 2008 European

Union Code of Conduct and similar recent initiatives offer recommendations without possessing the necessary legal mechanisms for enforcement.

The Treaty on Principles Governing the Activities of States in the Exploration and Use of Outer Space, Including the Moon, and Other Celestial Bodies, established in 1967, offers a comprehensive legal foundation for the peaceful utilisation of outer space. It serves as a fundamental framework for the progressive development of space law. The Space Accords have been ratified by numerous governments, and a larger number of governments adhere to the principles outlined in these accords. Given the significance of international collaboration in the establishment of space law norms and its crucial role in fostering global cooperation for the peaceful utilisation of outer space. The objective of this summary is to consolidate the five sets of principles and the five outer space treaties that have been officially ratified by the United Nations into a comprehensive single volume (United Nations. Office for Outer Space Affairs. & United Nations. General Assembly, 2002).

2 Satellite Technology and Operations

There exists a diverse range of satellite orbital configurations and sizes. The most fundamental orbits in the field of aerospace engineering include geosynchronous or geostationary orbits (GEO), polar orbits, and low, medium, and high Earth orbits (LEO, MEO, and HEO). The utilisation of High Earth Orbit (HEO) is more widespread in the context of early-warning satellites, whereas Low Earth Orbit (LEO) is considered to be a more economically efficient option for Earth observation purposes. GPS satellites are situated within the MEO (Medium Earth Orbit) region. The estimated lifespan of a satellite is approximately 15 years. It is also more advantageous to position a non-operational satellite in an elevated "graveyard orbit" or to facilitate its controlled re-entry into the Earth's atmosphere, resulting in its incineration (Fritz, 2013).

The definition of satellite types is critical for understanding their vulnerabilities and potential cyber-attacks. Satellites are classified into several categories, where each performs a distinct role as follows Table 2.

The next aspect, in terms of criticality, is to understand the risk factors associated with space exploration and satellite technology, and this awareness extends to the emerging role of Unmanned Aerial Vehicles (UAVs) and drones, which have become integral components in securing communications. UAVs and drones are unmanned vehicles that have evolved to play a significant role in various domains, including military, civil, commercial, and scientific applications. In the context of securing communications, these unmanned vehicles offer unique advantages and face distinct challenges, which are important for protecting data and preventing attacks. Sharma et al., (2020), discusses four types of attacks and their countermeasures:

Table 2 Satellite types and orbit functions

Types of satellites	Altitude	Applications
Low-Earth Orbit (LEO) Satellites	Approximately 160 km to 2000 km	Earth observation, scientific research, some communication purposes. Lower latency communication
Medium-Earth Orbit (MEO) Satellites	2000 km to 35,786 km	Global navigation systems (e.g., GPS)
Geostationary Orbit (GEO) Satellites	Approximately 35,786 km	Television broadcasting, weather monitoring, constant coverage applications

- **Electronic fence attack:** An amateur drone invades a protected area by crossing a virtual boundary. The countermeasure is to use dual-frequency PIFA, directional antennas, and angle reflectors to detect and block the drone.
- **Jelly-fish attack:** A malicious node sends false control messages and causes network congestion in a mobile ad-hoc network (MANET). The countermeasure is to use a secure and efficient routing protocol based on trust and energy levels.
- **Encryption attack:** An adversary eavesdrops or spoofs the data transmission and authentication between UAVs and ground stations. The countermeasure is to use various encryption schemes, such as lightweight stream ciphers, elliptic curve cryptography, homomorphic encryption, and quantum key distribution.
- **Physical intrusion attack:** An adversary captures or destroys a UAV by physical means. The countermeasure is to use several defence mechanisms, such as RF jamming, 3D MIMO radar, and self-destructing UAVs.

Drones and UAVs can also be vulnerable to various attacks that can compromise their functionality and data integrity. Some of the common attacks are Denial of Service (DoS) attacks, which prevent the drones from communicating with the ground station or other drones, Man-in-the-middle attacks, which intercept and alter the data transmitted by the drones, and De-Authentication attacks, which disconnect the drones from the network (Hassija et al., 2021). To counter these threats, the research paper proposes using Blockchain, Machine Learning, Software Defined Networks, and Fog/Edge computing. These technologies can enhance the security, reliability, and efficiency of the drone network by providing cryptographic data storage, malicious drone detection, network traffic control, and distributed computation (Hassija et al., 2021).

Hashesh et al., (2022), showcase how artificial intelligence (AI) techniques can improve the performance and functionality of unmanned aerial vehicles (UAVs) communications for future wireless networks, they also present some ML approaches that can be used for UAV networks, such as bandits, federated learning, meta-learning, and how these approaches can help solve problems such as resource allocation, path planning, network optimization, etc., for UAV networks (Hashesh et al., 2022).

UAVs and drones have a wide range of applications in securing communications and providing various civil services. They can serve as communication relays, enhancing connectivity in areas with limited infrastructure, as well as play a crucial role in military and national security operations with advanced communication systems. Moreover, they can be used for monitoring, wireless coverage, remote sensing, delivery, and other civil purposes, which can reduce human risks and operational costs. However, these applications also pose several challenges for UAVs and drones, such as charging, collision avoidance, swarming, networking, and security. These challenges require novel solutions to ensure the safety, efficiency, and reliability of UAVs and drones (Yaacoub et al., 2020) and (Shakhatreh et al., 2019).

Despite their advantages, UAVs and drones face challenges and limitations in securing communications. Cybersecurity threats, like those affecting satellites, pose risks to the data transmitted by these unmanned vehicles. Ensuring the integrity and confidentiality of communication signals is a significant challenge. Moreover, regulatory and ethical considerations, such as privacy concerns and compliance with airspace regulations, present hurdles. The potential for jamming or interception of communication signals adds another layer of complexity. Space debris management becomes relevant, as improperly operated UAVs and drones could contribute to cluttering Earth's orbit. UAVs and drones play an increasingly vital role in securing communications alongside satellites, and having a comprehensive understanding of the associated risks is imperative. Balancing their applications and benefits with the challenges and limitations requires careful consideration, emphasizing the need for robust governance policies and frameworks to ensure the responsible and secure use of these unmanned vehicles in the dynamic space environment.

The dynamic and integrated relationship between the different types of themes within space exploration, satellite technology and use of drones and UAVs is shown in Fig. 1.

2.1 *Ground-Based Satellite Systems*

The core of satellite systems is made up of ground-based satellite operations, which include a variety of crucial tasks that make it possible to command, monitor, and communicate with satellites in orbit. The key component for assessing and managing the cybersecurity risks and threats connected with ground-based satellite operations is understanding the many elements, functions, and operational environments of these activities. By considering these elements, an efficient cybersecurity framework could be created that is specific to the needs and difficulties of ground-based satellite operations. The definition, types, important components, functions, and operating environment of ground-based satellite operations are explained in this section.

Ground-based satellite operations refer to the set of activities and infrastructure involved in managing and supporting satellite systems from Earth. These operations are essential for establishing communication links, collecting, and processing data, and ensuring the overall functionality and performance of satellites. Ground-based operations can be classified into several types based on their primary functions:

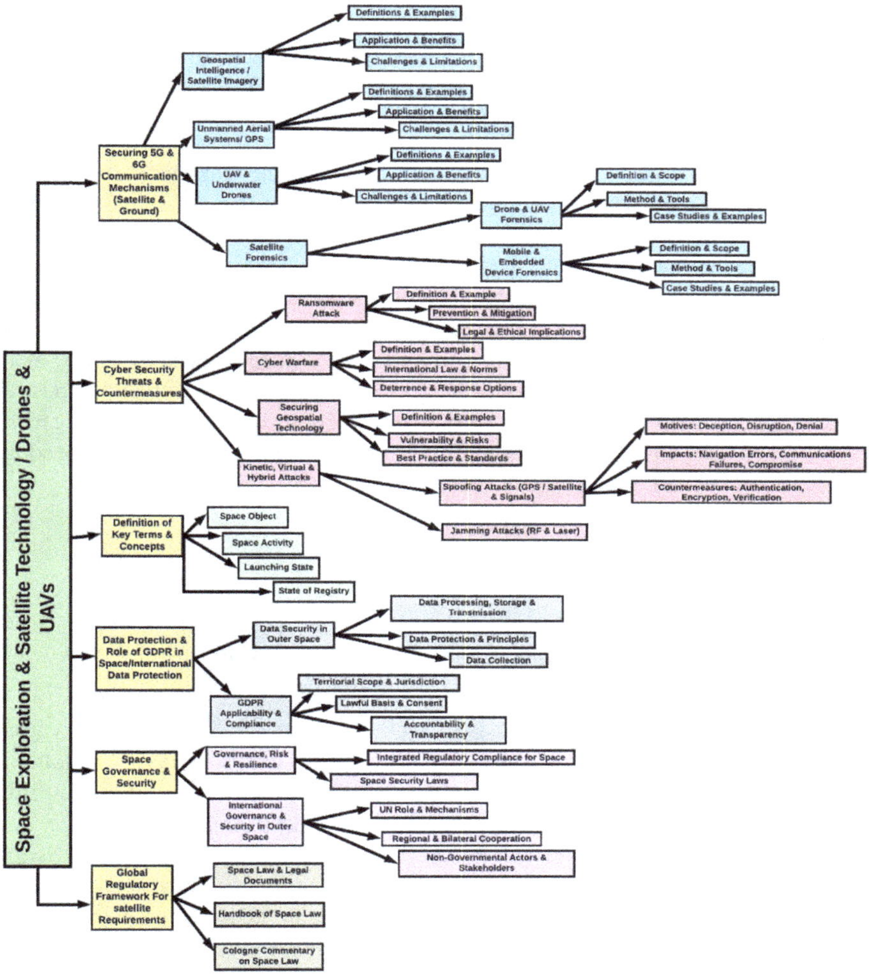

Fig. 1 Relationship between themes in space exploration, satellite technology and use of drones and UAVs

- **Telemetry, Tracking, and Command (TT&C):** TT&C operations involve monitoring and controlling satellites health, position, and orbital parameters. They include tasks such as sending commands to adjust satellite configurations, tracking their movements, and receiving telemetry data on satellite status. One of the challenges of satellite operations is the commissioning phase, which involves validating the health and functionality of the satellite subsystems after launch. According to Manulis et al. (2020), commissioning typically takes about two months and requires constant monitoring and control by the ground segment using (TT&C) systems.

- **Payload Operations:** Payload operations focus on the utilization of satellite payloads, which can include tasks such as capturing and processing remote sensing imagery, transmitting telecommunications signals, or conducting scientific experiments. These operations involve coordinating data collection, processing, and dissemination activities (Kodheli et al., 2021)
- **Network Operations:** Network operations pertain to the management and maintenance of the ground-based communication infrastructure that facilitates satellite connectivity. This includes managing ground stations, coordinating data transmission and reception, and ensuring network availability and reliability. According to Kodheli et al. (2021), satellite communications can benefit from ML/AI techniques in various ways, such as reducing interference, optimizing radio resources and network operations, and managing large satellite constellations. However, there trade-offs and benefits of ML/AI techniques over other conventional methods for satellite communications, such as optimization algorithms, signal processing techniques, etc., which may have different advantages and disadvantages depending on the context and objectives.

2.1.1 Key Components:

Ground-based satellite operations involve several key components and perform various critical functions to support satellite systems. The following components and functions are integral to the operations:

- **Ground Control Center (GCC):** The GCC serves as the central hub for managing and controlling satellite operations. It houses the necessary infrastructure, hardware, and software systems to communicate with satellites, monitor their health, and execute commands. The GCC also facilitates data processing, storage, and analysis. One of the potential threats to the security of satellite systems is the unauthorized access to ground control centre, which are responsible for communicating with and controlling the satellites. This could be achieved by exploiting poorly configured or vulnerable technologies, such as outdated software or weak encryption, or by phishing, which is a technique of deceiving users into revealing their credentials or installing malware (Manulis et al., 2020) This scenario is similar to the attacks on enterprise IT networks, which also rely on exploiting technical or human weaknesses to compromise the system (Manulis et al., 2020).
- **Ground Stations:** Ground stations are strategically located facilities equipped with antennas and receivers to establish communication links with satellites. They are responsible for transmitting commands, receiving telemetry data, and exchanging information between satellites and the GCC. Ground stations are often distributed geographically to ensure continuous coverage and efficient data transfer. Pavur and Martinovic (2022) assume that satellite ground stations are not different from any other terrestrial computing network, but they do not provide any evidence or justification for this claim or what these similarities, how they affect the performance or security of the satellite network.

- **Data Processing and Analysis Systems:** These systems are responsible for receiving, processing, and analysing the data collected by satellites. They employ various algorithms, software tools, and databases to extract meaningful information from the collected data, perform quality checks, and generate actionable insights. Shared ground station networks, such as those offered by cloud-based services (e.g., Amazon Web Services), can facilitate the communication and data processing of different constellations by providing access to high performance computing resources (Kodheli et al., 2021).
- **Command and Control Interfaces:** Command and control interfaces enable operators to interact with satellites, send commands, and receive telemetry data. These interfaces provide a user-friendly environment to monitor satellite status, configure parameters, and execute operational tasks.

2.1.2 Description of the Operational Environment

Ground-based satellite operations are conducted in a complex operational environment that encompasses technological, physical, and organizational aspects. This environment includes:

- **Technological Infrastructure:** Ground-based satellite operations rely on advanced technologies such as satellite communication protocols, data encryption mechanisms, network infrastructure, and software systems. The operational environment must address technological considerations to ensure secure and efficient communication and data management. Tsochev et al. (2020) claims that a mix of physical and technological measures, as well as consumer education, is necessary to prevent the attacks. However, they do not provide sufficient evidence or examples to support this assertion and present questions. How effective are these methods in different contexts and scenarios? What are the potential drawbacks or limitations of relying on them? How do they address the root causes of the attacks or the motivations of the attackers?
- **Physical Infrastructure:** Ground-based operations require dedicated facilities, including the GCC, ground stations, power and cooling systems, and physical security measures. The physical infrastructure must be designed to withstand environmental factors, provide redundancy, and protect against unauthorized access. According to Tosun (2021) the vulnerability of vital infrastructure to terrorist attacks poses a serious challenge to national security and social stability. Terrorists aim to cause maximum damage, disruption, and demoralization by targeting critical systems and services that support the country's economy and public welfare (Tosun, 2021).
- **Organizational Structure:** Ground-based satellite operations involve a structured organizational setup with specialized roles and responsibilities. This may include operators, engineers, cybersecurity professionals, and management personnel. The operational environment should promote effective coordination,

communication, and collaboration among different stakeholders. According to Li and Liu (2021) a new trend in organizational management is to hire a senior data security manager or a senior manager who oversees the various aspects of the security status of organizations.

2.2 Satellite Communication Protocols and Standards

Satellite communication protocols and standards play a crucial role in ensuring efficient and secure data exchange between ground-based stations and orbiting satellites. These protocols define the rules and conventions for data formatting, transmission, error detection, and correction, while standards provide a framework for interoperability and compatibility among different satellite communication systems. Some of the common communication protocols used in Ground-Based Satellite operations are listed below:

- **CCSDS Protocols (Consultative Committee for Space Data Systems):** CCSDS (2015) has developed a set of standardized protocols widely used in space missions, including satellite communications. Among these protocols, the CCSDS Telemetry, Tracking, and Command (TM/TC) protocol is particularly critical for exchanging telemetry data and commanding satellite operations. CCSDS File Delivery Protocol (CFDP) is another significant protocol used for reliable file transfer between space systems.
- **SCPS Protocols (Space Communications Protocol Standards):** The SCPS protocols provide a suite of optimized communication protocols tailored for satellite networks. These protocols aim to enhance the efficiency of data transmission over long-delay communication links, making them ideal for space-based operations.
- **DVB-S2 (Digital Video Broadcasting - Satellite - Second Generation):** DVB-S2 is a widely adopted standard for digital satellite communication, especially for broadcasting multimedia content. It offers improved spectral efficiency, higher data rates, and better error correction compared to its predecessor, DVB-S.
- **TCP/IP over Satellite:** Although TCP/IP is primarily designed for terrestrial networks, its adaptation for satellite communication is common. However, the long propagation delays in satellite links can lead to degraded performance. Various optimization techniques, such as TCP acceleration and proxy-based approaches, have been employed to address these challenges.

As technology continues to evolve, satellite communication protocols and standards will also undergo refinements and updates to meet the growing demands for higher data rates, improved security, and seamless interoperability. The Standards listed below look to secure the satellite communications.

- **AES Encryption (Advanced Encryption Standard):** AES is widely recognized as a secure encryption standard used to protect sensitive data transmitted over satellite links. It provides strong encryption algorithms that safeguard the confidentiality and integrity of data.
- **IPSec (Internet Protocol Security):** IPSec is a suite of protocols used to secure internet communications, including those over satellite links. By encrypting data at the IP layer, IPSec ensures secure end-to-end communication between satellite ground stations and user terminals.
- **SSPA/HPA Standards (Solid State Power Amplifier/High-Power Amplifier):** These standards define the requirements and performance characteristics of power amplifiers used in satellite communication systems. Ensuring compliance with these standards helps maintain reliable and interference-free satellite communications.
- **Viterbi and Reed-Solomon Coding:** Error correction coding schemes like Viterbi and Reed-Solomon are employed in satellite communication to combat signal degradation caused by atmospheric and space-related interference. These coding techniques enable the recovery of lost or corrupted data during transmission.

Satellite communication security is a field that is always changing due to technological improvements and the appearance of fresh dangers (Hosseinidehaj et al., 2019). Anomaly detection systems, software-defined networking for dynamic security setups, and quantum-resistant encryption are a few promising technologies that might improve the security of satellite communication against knowledgeable adversaries.

2.3 Satellite Threats

Cyber-attacks can target many aspects of a satellite system's functions, such as its ground systems, space units, and links between them. The easiest approach to interfering with a satellite is eventually to compromise the ground station since it offers the tools and software needed to properly operate and track it, as well as makes use of tried-and-true terrestrial systems and attack vectors (Manulis et al., 2020). Satellites and their ground control systems are essential for various industries and critical infrastructure, but also vulnerable to cyber-attacks. There are two main reasons why space systems are attractive targets for hackers: they are central points of failure that can cause widespread damage or disruption, and they lack common cybersecurity standards and regulation. The complexity of the supply chain and the multiple vendors involved in creating and operating space systems also increases the opportunities for hackers to compromise them (Falco, 2018).

The incorporation of artificial intelligence (AI) into cyber attacks marked the start of a new age of risks to satellite systems and has greatly increased the threat. The flexibility, automation, and sophistication of these attacks differentiate them. Understanding the growing threat environment posed by AI-driven attacks is critical for creating robust security measures to secure satellite infrastructure and assure the continuation of critical services.

- **Data Manipulation:** AI can be used to manipulate data transmitted by satellites. Attackers can alter or inject false information into satellite communications, leading to misinformation, navigation errors, or compromised data integrity.
- **Jamming and Interference:** AI-powered jammers can dynamically adjust their signals to disrupt satellite communication links. These jammers can target specific frequencies, making it challenging to mitigate interference.
- **Advanced Social Engineering:** AI can enhance social engineering attacks, enabling attackers to craft highly convincing phishing emails or messages. This can lead to unauthorized access to satellite systems or compromise sensitive information.

According to Falco (2018), public space asset organizations, like NASA, have taken several steps to improve their security, including implementing stricter access control policies, forming specialized teams for mission system security, encrypting data and communication networks, and collaborating with security researchers and experts. Through games and awareness initiatives, they have also built a security culture among the company's employees. However, they continue to encounter a wide variety of threats and challenges, including complicated supply chains, various merchants, outdated systems, and a lack of common standards.

Due to concerns about competition, reputation, or liability, private space asset firms such as SpaceX and Virgin Galactic are frequently less transparent and accessible to security researchers. They may also fail or delay addressing vulnerabilities revealed by ethical hackers or penetration testers. They do not have to disclose or report their cybersecurity testing procedures or results to the public or any governing body. This makes it difficult to evaluate and improve their security posture compared to other critical infrastructure sectors(Falco, 2018). It is essential to examine the efforts and challenges encountered by both the public space asset organizations, such as NASA, and private space asset companies like SpaceX or Virgin Galactic.

Publicly funded space agencies, exemplified by NASA, have recognized the critical importance of cybersecurity for safeguarding their space assets. These organizations have taken a number of proactive steps, such as:

- **Stricter Access Control Policies**: NASA and similar agencies have implemented stringent access control policies, limiting access to mission-critical systems to authorized personnel only. This includes measures such as two-factor authentication and biometric access controls.
- **Specialized Security Teams:** Dedicated teams specializing in mission system security have been established. These teams focus on monitoring, detecting, and

mitigating cybersecurity threats specific to space missions, ensuring that satellite systems remain secure throughout their operations.

- **Encryption of Data and Communication Networks:** Encryption protocols are deployed to protect the confidentiality and integrity of data transmitted to and from satellites. This prevents eavesdropping and unauthorized access during data transmission.
- **Engagement with Security Experts:** Public space asset organizations actively engage with security researchers and experts. This collaboration allows for the identification of vulnerabilities and the development of robust security measures. They also encourage responsible disclosure of security flaws.
- **Fostering a Security Culture:** NASA and similar entities have fostered a security culture among their employees. This includes awareness campaigns, training, and even gamification to educate staff about cybersecurity best practices and the importance of vigilant security measures.

However, despite these commendable efforts, public space asset organizations continue to confront various threats and challenges. Issues such as complex supply chains, multiple vendors, legacy systems, and the absence of common standards and regulations, present ongoing security concerns. These challenges underscore the need for continuous adaptation and innovation in satellite cybersecurity strategies.

In contrast, private space asset companies, including industry giants like SpaceX and Virgin Galactic, face distinct cybersecurity dynamics. Concerns related to competition, reputation, and liability often lead to different approaches, such as:

Transparency Challenges: Private companies tend to be less transparent and accessible to security researchers due to competitive considerations. They may limit access to their systems and data for fear of exposing proprietary information.
Response to Vulnerabilities: Some private space asset companies may ignore or delay addressing vulnerabilities reported by ethical hackers or penetration testers. The motivation behind this reluctance can stem from concerns about potential damage to their reputation or a fear of legal liabilities.
Lack of Public Reporting: Unlike public organizations, private companies are not obligated to disclose or report their cybersecurity testing procedures or results to the public or any governing body. This lack of transparency makes it challenging to evaluate their security posture, particularly when compared to other critical infrastructure sectors.

These dynamics and approaches highlight various issues associated with safeguarding private space assets. Balancing security and competitiveness, as well as preserving public trust is a constant challenge for these organisations.

2.3.1 Threats and Vulnerabilities to Ground-Based Satellite Systems

Cybersecurity risks and threats pose significant challenges to the integrity and security of ground-based satellite operations. The security of satellite-to-satellite communication links is a crucial aspect for the reliability and availability of satellite-based communication systems. However, most of the existing security solutions for SATCOMs focus on the ground segment or the user terminals, neglecting the potential threats and vulnerabilities that affect the space segment. According to Tedeschi et al. (2022) this is due to the difficulty of obtaining information about the communication protocols used by satellite links, which are often proprietary and confidential, and to the assumption that such links are inherently secure and isolated from external attacks. However, this assumption is questionable, as satellite links are exposed to various types of attacks, such as jamming, spoofing, eavesdropping, and hijacking, which can compromise the integrity, confidentiality, and availability of the transmitted data. Moreover, the authors argue that most of the current satellite systems do not implement adequate security mechanisms or rely on outdated and ineffective techniques. Therefore, they propose a novel security framework for satellite-to-satellite communication links, based on quantum key distribution and physical layer security techniques. Cybersecurity risks and threats encompass a wide range of malicious activities that can exploit vulnerabilities in the information technology infrastructure supporting ground-based satellite operations.

The security of satellite-to-satellite communication links is a crucial aspect of the reliability and availability of satellite-based communication systems. However, most of the existing security solutions for SATCOMs focus on the ground segment or the user terminals, neglecting the potential threats and vulnerabilities that affect the space segment. According to Tedeschi et al. (2022), this is due to the difficulty of obtaining information about the communication protocols used by satellite links, which are often proprietary and confidential, and to the assumption that such links are inherently secure and isolated from external attacks. However, this assumption is questionable, as satellite links are exposed to various types of attacks, such as jamming, spoofing, eavesdropping, and hijacking, which can compromise the integrity, confidentiality, and availability of the transmitted data. Moreover, the authors argue that most of the current satellite systems do not implement adequate security mechanisms or rely on outdated and ineffective techniques. Therefore, they propose a novel security framework for satellite-to-satellite communication links, based on quantum key distribution and physical layer security techniques. Cybersecurity risks and threats encompass a wide range of malicious activities that can exploit vulnerabilities in the information technology infrastructure supporting ground-based satellite operations.

These risks and threats can be categorized into several key areas:

- **Ground-based satellite operations:** These face unique cyber threats due to their specific operational characteristics and requirements. It makes sense and has been thoroughly researched why anyone would want to harm satellites. Identifying and classifying these threats are essential for developing targeted mitigation strategies (Livingstone & Lewis, 2016). In terms of potential attackers, a 2016 report by Chatham House, a well-known UK policy think-tank, classifies security actors into four main groups: nations looking for an edge, organized crime operations

for financial gain, terrorist groups looking for recognition, and lone hackers showcasing their skills. The Consultative Committee for Space Data Systems (CCSDS) list of attackers may be utilised to complement this perspective.
- **Command and Control System Compromises:** Threats targeting the command-and-control systems used to communicate with satellites can disrupt or manipulate satellite operations. These threats may involve unauthorized command transmission, injection of malicious commands, or manipulation of telemetry data. Over the 1990s, there were several attacks on satellite ground stations. These were high-profile events when hackers allegedly gained control over systems that would have allowed them to command spacecraft in orbit. The most noteworthy of them is a situation that occurred in 1998 in which hackers, generally suspected to be connected to the Russian government, obtained access to flight control systems at NASA's Goddard Space Flight Centre (Fritz, 2013). Some others link these incursions to a subsequent event where the German-US ROSAT X-ray telescope changed its orientation to aim optical sensors straight at the sun, resulting in catastrophic hardware damage.
- **Ground Station Vulnerabilities:** Ground stations, being critical communication links with satellites, can be targeted through cyber threats. Vulnerabilities in ground station equipment, software, or network infrastructure can lead to unauthorized access, data interception, or disruption of communication. In 2016, GNSS equipment revenue accounted for $84.6 billion of the $113.4 billion in overall ground equipment revenue, which has been increasing steadily since 2012 (Bryce Space and Technology, 2017). If a physical attack is successful in exploiting a vulnerability, the ground station may be rendered inoperable, which would have an immediate impact on how the mission is carried out and the services offered. If it wants to seize control of the spaceship without really harming the systems, it can potentially try to capture the facility (NASA Cybersecurity, 2012). According to a NASA assessment, the theft of an unencrypted notebook computer resulted in the loss of command-and-control algorithms for the ISS (Manulis et al., 2020).
- **Supply Chain Attacks:** Ground-based satellite operations rely on a complex supply chain involving hardware, software, and services from various vendors. Cyber threats can exploit vulnerabilities in the supply chain, leading to compromised components, malicious software injections, or unauthorized access to sensitive systems. Authentic components could not be accessible due to supply chain disruptions, which could lead to the usage of counterfeit parts (CCSDS, 2015). If confidence is not established, fake equipment or software may be given and utilized covertly on a mission. Malicious circuits or malicious code might be included in the hardware or software, which could have unwanted mission effects. Sensitive materials may be exposed or made accessible by supply chain partners. Suppliers who are connected and have security flaws might be exploited and utilized to harm organizations, systems, and missions (CCSDS, 2015).
- **Insider Threats**: The involvement of personnel with privileged access to ground-based satellite operations introduces the risk of insider threats. These threats may involve intentional or unintentional actions by employees or contractors

that compromise system security, leak sensitive information, or disrupt operations. The most frequent subcategories of threat actors, according to Tsochev et al. (2020), include terrorist organisations, insiders of businesses, and cybercriminals. Since it has been shown that each set of agents employs very similar attack tools, identifying them makes sense from a practical standpoint.

2.4 Privatisation and the Proliferation of Satellites

In 2013, the global space economy received a substantial contribution of $314 billion from government and commercial space operations. The level of governmental activity is gradually decreasing, whereas commercial space activity within the private sector is experiencing a more rapid expansion. During the period spanning from 2012 to 2013, there was a notable increase of 7% in revenue generated from commercial space products and services. In comparison, the commercial infrastructure and support industries experienced a growth rate of 5%, while government expenditure witnessed a decline of 2%. Significant investments in space are made to support the infrastructure of communications, which encompasses telecommunications and broadcasting, as well as weather and geological monitoring, and defence. Therefore, it can be observed that commercial space development is surpassing government space initiatives. The legal obligations and liabilities of spacefaring nations are expanding in tandem with the emergence of commercial space endeavors undertaken by private corporations (Kaul, 2017).

In recent times, there has been a notable surge in the quantity of satellites present in the Earth's orbit. As of January 31, 2022, the number of different satellites in orbit was recorded to be 8261, (Geospatial World, 2023). This may be attributed to advancements in space technology, a decrease in the expenses associated with launching satellites, and an escalating need for services that rely on satellite infrastructure. The rapid increase in satellite launches by governments, commercial organisations, and international organisations has prompted concerns regarding space debris, orbital congestion, and potential security risks. An essential requirement for guaranteeing the secure and environmentally responsible utilisation of satellite technology within this particular context is the implementation of an information policy or framework. Instances of cyberattacks against satellites have indeed been documented, and it is anticipated that the frequency and efficacy of such threats would escalate in the future, owing to the allocation of resources towards cyberwarfare and the refinement of attack techniques (Bichler, 2015).

The number of satellites in Earth's orbit has been steadily increasing due to advancements in technology, decreasing costs associated with satellite launches and commercial opportunities. Satellites play a crucial role in the functioning of contemporary civilization, particularly as technological advancements progress rapidly and global interconnectivity continues to grow. The existence of satellite cyberattacks has been substantiated, and the allocation of greater resources towards the advancement of cyberwarfare and attack techniques suggests that these threats are anticipated to become more frequent and effective in the coming years. The exponential proliferation of space debris presents a significant impediment to forthcoming endeavours in space exploration, hence posing a substantial threat to the integrity and functionality of satellites. The present situation underscores the imperative for a comprehensive global framework to tackle this matter and offers useful insights into the pervasive absence of satellite governance systems. The development of an adaptable international system that operates within a polycentric framework and demonstrates the ability to adjust to a dynamic global environment is of utmost importance.

The pursuit of space exploration and utilisation has proven to be a costly undertaking; however, it has undeniably yielded a significantly enhanced comprehension of our planet, the solar system, and the marvels of the cosmos. Satellites in Earth's orbit exhibit a diverse array of sizes and shapes, serving a broad spectrum of functions and applications. Satellites of considerable size, exceeding a mass of 1000 kg, facilitate the acquisition of detailed Earth remote sensing data, enabling high-capacity communication capabilities, and support advanced scientific research. However, the process of designing and deploying such satellites incurs significant expenses and demands a substantial investment of time. The availability of microelectronics in large quantities and at affordable prices has facilitated the utilisation of a distinct technique, helped by the advancements in miniaturisation. As a consequence, the development of physically smaller satellites enables the provision of functional and commercially viable services, while concurrently leading to substantial reductions in timelines and prices. This study documents the historical development and increasing prominence of small satellites, tracing their origins as a subject of curiosity with limited practical utility to their current status as a crucial element of contemporary space capabilities, Sweeting, (2018).

The emergence and extensive utilisation of microelectronics have enabled the production of smaller satellites by smaller and agile teams with limited resources. These satellites can be launched as secondary payloads alongside larger, paying satellites, taking advantage of available launch capacity. This event resulted in a significant alteration of the power dynamics within the realm of space exploration. Consequently, the commencement of the modern small satellite-era transpired during the early 1980s. Undoubtedly, the deployment of diminutive spacecraft was not a novel concept. Numerous initial satellites launched by the United States and the Soviet Union, along with subsequent experimental satellites from various nations, can be classified within this category. Consequently, approximately 1500 satellites, weighing less than 100 kg, have been launched worldwide over the past six decades.

Contemporary small satellites have demonstrated the potential to rival, and occasionally surpass, the functionalities of conventional large-scale satellites while being significantly more cost-effective. The constellations commonly referred to as mega-constellations are expected to facilitate the advancement of small satellites and potentially contribute to the development of small launchers. Moreover, these advancements have the potential to facilitate the development of autonomous space assembly techniques for large-scale systems comprising a multitude of intelligent nano- and micro-satellite components, resembling "Lego" pieces. Additionally, they could also enable the in-orbit fabrication of satellites and spacecraft. These technological advancements have the potential to result in more than just the establishment of "digital factories" solely on Earth. Small satellites, similar to their terrestrial counterparts such as laptops and smartphones, have gained significant prominence and are facilitating the emergence of novel business models. The emerging 'NewSpace' sector exhibits a dynamic and innovative nature, holding significant potential to revolutionise the space industry and expand the range of services offered by space, ultimately benefiting the global population as a whole (Sweeting, 2018).

2.4.1 Satellite Constellations

Around two decades ago, Iridium and Globalstar emerged as pioneering satellite constellation concepts in the market. The expansion of access to space is being facilitated by advancements in technology that have led to the miniaturisation of space-related equipment and the development of sophisticated design capabilities. As a result of the significant projected surge in the frequency of small satellite launches, particularly those of pico to mini sizes, sustainable enterprises are redirecting their focus towards constellations. The communication industry exhibits significant potential for the implementation of constellations. In order to facilitate the forthcoming era of the Internet-of-Things, it is imperative that the communication infrastructure possesses the capacity to effectively manage substantial quantities of data and ensure uninterrupted service across all geographical areas. However, constellations also possess significant potential in the fields of meteorology, security, and disaster monitoring (Curzi et al., 2020).

2.4.2 Challenges of Space Debris

The proliferation of extensive constellations of small spacecraft poses a significant threat to the environment and safety in space, as it will inevitably result in a substantial escalation of space debris. The existing state of affairs presents a substantial challenge to the global management of space debris. The existing space debris mitigation standards exhibit limited adaptability and effectiveness when confronted with the challenges posed by the proliferation of numerous small spacecraft constellations in the contemporary era. Given the circumstances, it is imperative to establish an innovative framework for the management of space debris that encompasses measures for mitigating and preventing space debris, as well as prioritising active removal.

2.4.3 Management of Space Debris

The lack of a robust regulatory framework has led to a notable contribution of satellites towards the proliferation of space debris, which has become an increasingly disturbing issue. The presence of this debris has a notable adverse impact on the environmental conditions within Earth's orbit. According to estimates, around 2000 out of the 6000 satellites launched since the inception of Sputnik I in 1958 continue to orbit the Earth, albeit in a non-operational state. Presently, the number of trackable fragments of debris resulting from deliberate or inadvertent incidents exceeds 20,000. The regulation of contemporary issues such as satellite data security and orbital debris lacks a comprehensive framework that can effectively address these concerns. The Department of Defence (DoD) currently bears the responsibility of overseeing the monitoring of space objects; however, it lacks the requisite authority to mandate that satellite operators implement precautionary measures to mitigate the risk of potential collisions.

3 Cybersecurity AI-Based Attacks and Quantum Computing Impacts

Satellite technology plays a crucial role in various domains such as communication, worldwide broadcasting, early warning systems, navigation, reconnaissance, remote sensing, meteorology, and surveillance. Satellite systems are widely recognized as "critical infrastructure" by numerous governments, and any disruptions to these systems can result in severe and disastrous repercussions.

Artificial Intelligence (AI) is the science that studies how to create software that can carry out operations that normally need human intelligence. The two main subdisciplines of this interdisciplinary area, which draws on computer science, mathematics, logic, psychology, and other fields, are symbolic AI and sub-symbolic AI (Russell & Norvig, 2020). The history of AI goes back to the Greeks, who tried to use logic and mathematics to simulate human thought and behaviour. John McCarthy then first used the phrase artificial intelligence in 1955 at a symposium at Dartmouth College when he invited scientists from diverse disciplines to explore the potential and difficulties of building intelligent robots (McCarthy et al., 1995).

Since then, artificial intelligence has experienced several cycles of advancement and setbacks known as AI winters and AI springs. Some of the milestones to mention are the creation of the first AI programming languages (LISP and Prologue), the first expert systems (DENDRAL and MYCIN), the first chess program (Deep Blue) to defeat a human world champion (Garry Kasparov), the first conversational agents (ELIZA and ALICE), the first deep learning models (LeNet and AlexNet) Wooldridge (2020). Web search engines, recommendation systems, voice or handwriting recognition, self-driving cars, chatbots, computer vision, natural language processing, robotics, healthcare, education, entertainment, security, and other fields are just a few of the many fields where AI is being used (Tegmark, 2017).

The rapid advancement of artificial intelligence (AI), particularly in the realm of machine learning and deep learning, has introduced a transformative element in the landscape of cyber threats and attacks. AI has become a powerful tool for malicious actors seeking to exploit vulnerabilities, evade detection, and launch sophisticated cyber-attacks.

The following discusses the role of AI in cyber-attacks and its implications for satellite security in these AI-Driven attack techniques.

- **Adaptive Threats:** Cyberattacks powered by artificial intelligence are flexible and dynamic. Machine learning algorithms are used by attackers to constantly alter their attack strategies in response to changes in the security environment. Traditional defensive systems are having difficulties keeping up with these developments.
- **Automated Exploitation:** Artificial intelligence-powered attacks can automate the process of discovering and exploiting vulnerabilities in satellite systems. Attackers utilise AI algorithms to find weaknesses, allowing them to perform large-scale assaults with minimal human involvement.
- **Stealth and Evasion:** Artificial intelligence could be used to develop enhanced evasion methods. Attackers, for example, can utilize AI to create malware that is capable of altering its behaviour to avoid detection by typical antivirus software or intrusion detection systems.
- **Zero-Day Exploits:** AI could be used for discovering and exploiting zero-day vulnerabilities, which are not previously recognized security weaknesses. This provides attackers with a substantial edge when it comes to conducting targeted assaults.

There are also consequences on Satellite Systems in the following techniques:

- **Data Manipulation:** AI can be used to manipulate data transmitted by satellites. Attackers can alter or inject false information into satellite communications, leading to misinformation, navigation errors, or compromised data integrity.
- **Jamming and Interference:** AI-powered jammers can dynamically adjust their signals to disrupt satellite communication links. These jammers can target specific frequencies, making it challenging to mitigate interference.
- **Advanced Social Engineering:** AI can enhance social engineering attacks, enabling attackers to craft highly convincing phishing emails or messages. This can lead to unauthorized access to satellite systems or compromise of sensitive information.

Understanding the AI-driven threats to satellite systems is essential for developing effective countermeasures. Some approaches include:

- **AI-Powered Defence:** Using AI and machine learning to detect anomalies and identify threats in satellite networks.
- **Behaviour Analysis:** Observing the behaviour of satellite systems and users in order to discover abnormal trends that may indicate the possibility of an attack.
- **Security Patching:** Patching vulnerabilities promptly to limit the risk of zero-day exploits.

The incorporation of artificial intelligence (AI) into cyber attacks marked the start of a new age of risks to satellite systems. The flexibility, automation, and sophistication of these attacks differentiate them. Understanding the growing threat environment posed by AI-driven attacks is critical for creating robust security measures to secure satellite infrastructure and assure the continuation of critical services.

AI-based cyber-attacks use machine learning algorithms, data poisoning, model theft, or adversarial samples to enhance their capabilities and evade detection. The impact of these attacks on networks can be severe, such as signal interference, loss of control, and permanent damage to network hardware or software. Hackers can use data poisoning to compromise machine learning models used for network image analysis or navigation (Manesh & Kaabouch, 2019). They can also use model theft to gain access to proprietary or confidential machine learning models used by network operators or users. Adversarial samples can be used to deceive or mislead network systems or networks that rely on machine learning models for decision-making or classification.

Some major security concerns associated with AI-based cyber-attacks on networks include:

- **Compromised Communication:** Hackers can use various techniques to compromise communication between network and ground stations, such as injecting malware or hijacking the link segment. Such attacks can result in data breaches, loss of control, or interference with network signals, affecting the overall functioning of network systems.
- **Unauthorized Access:** AI-based cyber-attacks can be used to gain unauthorized access to network systems, including sensitive data or proprietary machine learning models. This can lead to significant security breaches, resulting in damage to the network system and its users.
- **Machine Learning Model Poisoning:** Hackers can use data poisoning techniques to manipulate machine learning models used in network systems. Such attacks can cause incorrect predictions, leading to erroneous decision-making and affecting the overall functioning of network systems (Chachra & Sharma, 2019).
- **Signal Jamming and Interference:** Network signals can be jammed or spoofed by hackers to interfere with services related to communication and navigation. Such assaults may have an effect on how well the overall network system operates, which may damage its users.

Therefore, it is essential to create efficient techniques to deal with AI-based cyber-attacks since they represent serious security problems for network systems. The proposed strategies, including AI-based defence mechanisms, enhanced cybersecurity protocols, and collaborative efforts, aim to safeguard the integrity and functionality of network systems. There needs to be a higher and more accelerated call to action for the network industry and cybersecurity community to collaborate and develop effective strategies to address these emerging AI-driven threats.

3.1 Impacts of Quantum Computing on Satellite Communications

The provision of dependable and secure communication lines between ground stations and orbiting satellites has advanced significantly in the field of satellite communication. Numerous crucial applications; such as communications, Earth observation, and national defence, depend on these systems. However, the development of quantum computing technology poses fresh difficulties that risk upending established secure communication paradigms. The following examines the security of satellite communications as it is now and evaluates how susceptible it is to risks from quantum computing. Techniques for encryption are used in satellite communication to protect the privacy and reliability of information sent. These encryption techniques have progressed, and contemporary systems now use cutting-edge encryption techniques like RSA and ECC (Elliptic Curve Cryptography) to safeguard sensitive data, according to Thomas, A. and Manuel, E.M. (2016). Although these encryption techniques have proven resistant to classical computers, the situation changes when quantum computers are involved.

Quantum computers have the potential to hack widely used encryption algorithms by exploiting their inherent vulnerabilities to quantum attacks. The most notorious among these attacks is Shor's algorithm, which can efficiently factor large integers and solve the discrete logarithm problem. Quantum computers are still in their infancy and face many technical challenges before they can achieve the level of performance and reliability required to run Shor's algorithm effectively. Moreover, there are alternative encryption algorithms that are resistant to quantum attacks, such as lattice-based cryptography and hash-based signatures. These algorithms can be used to protect satellite communication data from quantum adversaries. These two mathematical problems underlie the security of many encryption algorithms, including RSA and ECC. As quantum computers advance in their capabilities, the timeframe within which they can potentially crack encryption keys shortens. What was once considered a secure communication method against classical computers may become obsolete, raising concerns about the confidentiality of satellite communication data. What steps should business executives take as quantum technologies are ready to make a splash in the commercial world?

Researchers and organisations are working to create quantum-resistant cryptography methods in response to the looming threat of quantum computing. These algorithms are meant to survive quantum computer assaults and preserve data security in a post- quantum environment. There is a growing understanding among those involved in satellite communication of the necessity of switching to quantum-resistant encryption techniques. For example, to protect satellite operations, the Consultative Committee for Space Data Systems (CCSDS) is working hard to build quantum-resistant communication protocols.

Although post-quantum cryptography is advancing, applying quantum-resistant security measures in satellite communication systems poses difficulties. Satellites frequently operate in resource-constrained situations and have limited computing resources. Thus, while implementing new cryptographic algorithms, it is important to consider how much energy and computing overhead would be required. Furthermore, amongst satellite operators, ground stations, and communication equipment makers, careful planning and cooperation are required to make the shift to quantum-resistant encryption. This shift may entail hardware upgrades and adjustments to operational procedures; it is not only a software update.

The current state of satellite communication security is robust against classical computing threats. However, the impending era of quantum computing introduces a new dimension of vulnerability. To ensure the continued confidentiality and integrity of satellite communication data, the satellite industry must proactively adopt quantum-resistant encryption methods and address the unique challenges associated with their implementation in satellite systems which is going to be proposed in this framework. This transition will be pivotal in safeguarding the future of satellite communication in the quantum computing era.

4 The Road to an Effective Space Governance Framework and Cybersecurity Mitigations

Satellite communications have been of paramount importance in the daily lives of individuals since its establishment 65 years ago. These technologies facilitate the reception of television and radio broadcasts, the transmission of crucial communications to isolated terrestrial regions, maritime transportation, and the observation and anticipation of environmental conditions. Furthermore, they facilitate individuals in actively assuming positions and navigating. Indeed, significant advancements have been made in the realm of satellite communications during recent decades.

Despite significant advances in space technology, the current international framework governing space operations is profoundly influenced by the geopolitical climate of the Cold War. The primary objective of the architects of space law during that era was to establish measures that would ensure the peaceful use of outer space and prevent the expansion of the Cold War into this domain. From a more sceptical vantage point, it could be argued that both superpowers demonstrated a willingness to employ a legally enforceable agreement in order to reduce the likelihood of the other party attaining a significant military advantage in space. Numerous aspects of the legal framework established by these accords in relation to satellite regulation require additional analysis, taking into account national law. Satellites have a significant effect on the production of space debris, but the absence of a stringent regulatory framework has increased concern in this regard. The presence of this particular type of human-made waste in Earth's orbit causes severe environmental damage.

Therefore, any object launched into space, such as malfunctioning satellites or those that have reached the end of their useful existence, has the potential to become space debris. The Earth's atmosphere has been subject to a continuous and uninterrupted presence of space-related operations for more than five decades. A significant proportion of the roughly 6000 satellites launched since the launch of Sputnik I in 1958, namely more than 2000, continue to occupy orbital space despite having lost their operational functionality. There are already over 20,000 identifiable debris fragments that can be tracked, the consequence of both intentional and unintentional collisions. Due to their exceptionally high impact velocities while in orbit, such debris is extremely hazardous. Significant velocities can have devastating effects on operational satellites, as tiny particles can cause calamitous damage. Nonetheless, the current pervasiveness of anthropogenic pollution poses a significant obstacle to the planet's ability to address this issue. The current legal framework for debris management is a complex amalgamation of informal, non-binding norms and additional components of international space law.

Due to the developments in the Ariane 6 rocket and the competitive dynamics between Space X and Blue Origin, which are allocating significant resources to the development of reusable rockets, the profitability of the commercial launch sector continues. As stated previously, there is a tension between the imperative for development and the potential threats to space security and the environment. This disagreement poses significant challenges to institutions and the existing legal and regulatory framework. States and businesses who disregard this threat do so at their own peril. To ensure the viability of human space activities, it is necessary to revive the regulatory framework for all satellites, which is as important as any technological or engineering advancement (Newman, 2017).

The United Kingdom has a strong inclination towards active involvement in the global space sector. The establishment of international satellite deployment stations is anticipated to lead to a further augmentation in the quantity of satellites. Consequently, the global space community will prioritise the development of suitable strategies and policy mandates pertaining to cybersecurity across nations with regards to satellite communications. This will encompass considerations for both new and small satellites, as well as those already in operation. The United Kingdom's efforts to enhance its projects and capabilities in the field of satellite deployment and communication on the global stage could potentially be advantageous through the investigation of intergovernmental "asset pooling" among partners in the space community. The limited extent of research conducted on the numerous intergovernmental agreements that now exist is a subject of interest among the United Kingdom's space industry and academic community. In order to contemporize the Treaty and facilitate the sustained advancement of space exploration, it is imperative to recognise the inherent interdependence between scientific research and the essential implementation of cyber security measures. The sole regulatory framework now in effect consists of Memorandums of Understandings (MoUs), which are enforced by the asset owners. The current exploration of space by commercial private firms has brought attention to the unregulated nature of resource exploitation in space, particularly in relation to societal, ethical, and political considerations (Turner & Jahankhani, 2021).

Accountability in international space law is upheld by the involvement of state-owned bodies, including the International Telecommunications Satellite Organization (INTELSAT), European Space Agency (EUTELSAT), International Maritime Satellite Organization (INMARSAT), and Intergovernmental Organizations (IGOs).

These entities assume responsibility for overseeing all operations conducted within the realm of outer space. These companies exhibit a commercial nature, prioritizing operational and commercial concerns over regulatory matters (Newman, 2017).

The utilization of space-based solutions is increasingly gaining traction among both governmental and private entities, resulting in the overcrowding of the radio frequency spectrum and orbital slots. With the increasing accessibility of satellite technology for commercial purposes, there is expected to be a rise in the quantity of privately owned and operated satellite spacecraft in low-Earth orbit. These spacecrafts will be employed for the purpose of gathering data, facilitating access for government agencies, and providing affordable global internet connectivity. The satellite ground systems of the Air Force and the Department of Defense are often inadequately assessed for cyber vulnerabilities, resulting in their susceptibility to hacking and other forms of cyber threats. In order to effectively identify, assess, mitigate, and manage sophisticated cyberthreats, a consistent strategy to cybersecurity assessment is required (Bichler, 2015).

Satellites have the potential to induce environmental degradation within Earth's orbit, hence exacerbating the issue of space debris production. Due to the absence of a global regulatory framework governing satellite broadcasts, operators of satellite services may resort to employing clandestine tactics of intimidation, coercion, and persuasion.

The combination of the COBIT 2019 and ITIL 4 frameworks in satellite communications offers distinct advantages, as it ensures comprehensive consideration of all aspects of governance and administration. To instil confidence among international stakeholders on user privacy, integrity, accessibility, accountability, and non-repudiation (CIAAN) as well as compliance with the General Data Protection Regulation (GDPR), it is imperative to adopt a reduced risk posture (Turner & Jahankhani, 2021).

Satellites have significantly contributed to global affairs. Specifically, the Space Supersystem has heavily relied on its communications infrastructure. In the future, the utilization of satellites and space technology holds the potential to facilitate the establishment and operation of 5G and 6G networks. Private enterprises such as SpaceX aim to integrate a substantial number of 42,000 spacecraft into what is commonly referred to as a mega constellation. Presently, this constellation encompasses more than 4500 satellites of varying dimensions that orbit the Earth. The demand for satellite governance is driven by the diminishing dimensions and exponential proliferation of satellites within the Earth's orbit, Howell and Pultarova (2023). Satellite structures are susceptible to the presence of vulnerabilities that can be exploited in their nodes and linkages. The importance of satellite ground systems in cybersecurity is frequently disregarded in conversations about satellite security. The establishment of a global regulatory framework for satellites soon appears highly improbable, given the current absence of such a framework.

4.1 Comprehensive Cybersecurity and Overarching Space Asset Protection

The development of a comprehensive approach to effectively respond to and mitigate potential threats necessitates the implementation of a systematic and unified policy solution. This solution should effectively guide technological endeavours aimed at safeguarding space assets and services. The implementation of mechanisms to enforce rules is imperative in order to uphold the interests of legitimate users and actions, while simultaneously raising the costs associated with illegitimate users and actions. The emergence of novel actors, including states, non-state entities, and commercial entities, alongside advancements in technology, has brought about significant changes and transformations in space operations. Consequently, the formulation of policy responses necessitates addressing a diverse range of concerns. Currently, both space policy and cyber security policy are inadequately equipped to address the complex security implications arising from the convergence of cyberspace and outer space. This convergence has substantially amplified security concerns, necessitating a more robust and comprehensive approach to policymaking in these domains. The phenomenon of space commercialization, encompassing various market-driven mechanisms aimed at reducing costs, as well as entrepreneurial endeavours like asteroid extraction and space tourism, gives rise to apprehensions regarding cyber security. Furthermore, there is an emerging inclination towards the utilisation of diminutive satellite networks and novel satellite services across diverse sectors, including transportation, agriculture, and environmental monitoring. These advancements yield significant data that can be employed for espionage and unlawful endeavours.

The fundamental principle underlying the suggested policy solution is the incorporation of a comprehensive framework for cybersecurity, encompassing protective measures throughout all phases of space system development and operation. It is imperative to implement policies that enforce rigorous cybersecurity standards for all components of space systems and their supply chains, encompassing both civilian and military space assets and activities. One example that should be considered is the incorporation of the Cybersecurity Maturity Model Certification (CMMC) into the development of commercial space assets, as it has been mandated. The primary focus of the policy framework revolves around the significance of implementing suitable regulations for the commercial space industry. The implementation of a regulatory framework would offer a sense of regulatory assurance to commercial space enterprises, considering the projected growth of space activities undertaken by private entities. Simultaneously, this framework would facilitate states in fulfilling their obligations under any prevailing space treaties, such as the Outer Space Treaty. The integration of indicators and metrics within the policy framework is essential for assessing the actualization of anticipated policy effects (Varadharajan & Suri, 2022).

Cyber-physical systems, which are recent information and communication technology (ICT)-based systems, are consistently incorporating SCADA systems. The utilization of SCADA systems on satellites, which are classified as critical infrastructure, carries a significant vulnerability to ransomware attacks. Organizations must prioritize the inclusion of ransomware inside their risk management frameworks and incident response plans. The issue of cybersecurity in critical infrastructure is a matter of significant concern. Despite the implementation of robust security measures, there is a persistent vulnerability that allows malicious actors to illicitly infiltrate a system and exploit it for nefarious purposes (Ibarra et al., 2019).

4.2 Current Satellite Governance

The current global governance procedures pertaining to space debris prioritise the mitigation and prevention of such debris. Despite the presence of pertinent treaties, the regulations pertaining to the handling of space debris at an international level primarily reside within non-binding "soft law" documents, which nations have the option to adopt on a voluntary basis.

The existing framework of international space law has effectively established the fundamental legal structure for the governance of space debris. One example of such a treaty is the Outer Space Treaty, which serves to establish the essential legal norms for activities conducted in outer space. In the context of space debris management, this treaty underscores the significance of safeguarding the space environment through environmental protection measures. According to the provisions outlined in the Treaty, it is mandatory for all nations to carry out their space-related endeavours in adherence to the principle of environmental preservation, thereby abstaining from any actions that may result in the contamination of the Moon or any other celestial entity. The Interagency Space Debris Coordination Committee (IADC) adopted the Space Debris Mitigation Guidelines in 2002. The International Astronautical Development Consortium (IADC) is an intergovernmental organisation with a multilateral structure. Its primary objective is to facilitate international coordination and collaboration in addressing the issue of space debris on a global scale. The recommendations put forth by the International Association for the Advancement of Debris Mitigation (IADC) outline a comprehensive approach to managing space debris. These guidelines encompass four key aspects: the prevention of debris generation during regular operations, the reduction of on-orbit fragmentation, the removal of debris post-mission, and the prevention of on-orbit collisions. The guidelines for mitigating space debris were adopted by the Committee on the Peaceful Uses of Outer Space (UNCOPUOS) in 2007. The mitigation guidelines established by the United Nations Committee on the Peaceful Uses of Outer Space (UNCOPUOS) are derived from the International Association for the Advancement of Space Safety (IADC) standards mentioned earlier. These guidelines provide strategies for spacecraft to avoid potential collisions with space debris during different phases, considering the origins of such debris.

A comprehensive strategy for space debris management necessitates the deliberate consideration and integration of three key components: (1) proactive measures to prevent foreseeable collisions, (2) strategies for minimising the accumulation of orbital debris, and (3) efforts aimed at mitigating the existing orbital debris and more detailed as follows.

- **Prevent Future Space Debris:** The sustainability and security of space activities are significantly jeopardised by the presence of space debris, colloquially referred to as space junk. The emphasis is on three key domains: mitigation, tracking, and legal rules, to address the challenges posed by space debris. The implementation of these measures can effectively contribute to the enhancement of the cleanliness and safety of the space environment, thereby benefiting both ongoing and future space missions.
- **Minimise Orbital Space Debris Accumulation:** The mitigation of space debris is vital to prevent the accumulation of waste in the universe and is aimed at a multitude of measures for achieving effective mitigation of space debris. It is recommended to employ a comprehensive plan to enhance the sustainability of space activities and mitigate potential hazards and advisable to promote the utilisation of materials that undergo combustion during the re-entry phase into the Earth's atmosphere among satellite designers; since this practise can effectively decrease the likelihood of debris reaching the planet's surface. Furthermore, legislation must be enforced mandating spacecraft operators to adhere to either controlled re-entry protocols into the Earth's atmosphere or the transfer of dead satellites and upper stages to designated graveyard orbits that have been approved for such purposes. Furthermore, it is imperative to design procedures aimed at preventing collisions. These strategies should encompass the implementation of proactive space traffic surveillance and timely notifications in instances where satellites or space debris are at risk of collision. Lastly, a crucial element is the allocation of resources and the advancement of Active Debris Removal (ADR) technology to effectively retrieve and eliminate large, abandoned spacecraft and hazardous space debris that pose a threat to orbital trajectories. The implementation of these measures can contribute to the establishment of a more safe and enduring space environment.
- **Reduce Existing Space Debris**: The precise and prompt identification of space debris is crucial for establishing situational awareness in the realm of outer space. To enhance the monitoring capabilities of both small debris and celestial objects, the implementation of a comprehensive plan is required. To effectively monitor minute debris and ascertain precise positioning and predicted trajectories of celestial objects, there is a need for enhancements in ground-based radar systems. Space-based sensors can be effectively employed in conjunction with terrestrial radars to provide continuous monitoring of space debris from several perspectives. The promotion of international cooperation and data exchange across space agencies and organisations is vital for advancing these endeavours and enhancing the overall comprehensiveness and precision of space debris monitoring. The development of advanced data processing algorithms will play a significant role

in enhancing the safety and security of the space environment. These algorithms will enable the identification of probable collisions and the assessment of risks posed to space assets.

- **Space Debris Legislation:** To effectively address the issue of space debris, it is important to develop and enforce legal frameworks. There should be a comprehensive approach, which integrates domestic legislation and international agreements, be employed to address the issues pertaining to accountability and responsibility surrounding space debris. Federal regulations will be enacted to govern all space-related endeavours, encompassing the launching and operation of satellites, as well as the decommissioning of space assets once they have reached the end of their operational lifespan. Furthermore, there will be endeavours to provide assistance and uphold existing international accords, such as the Space Liability Convention and the Outer Space Treaty, which provide principles for addressing liability and regulating activities in outer space. The ultimate goal is to establish a globally coordinated framework for the management of space transportation, emphasising safety and ethical conduct in all activities pertaining to outer space. In order to promote adherence to debris mitigation requirements, a scheme of accolades will be instituted for enterprises and groups that demonstrate compliance, while punitive measures will be enforced upon those that fail to do so. The objective of this comprehensive strategy is to foster ethical conduct, mitigate the proliferation of space debris, and safeguard the integrity of the space environment for future generations.
- **Cybersecurity and Safety**: Satellite ground structures play a crucial role in space missions as they are responsible for the supervision of satellite operations and the secure transmission of sensitive information. The ground systems, which encompass the processes of data translation and transfer, play a pivotal role in both cyber and space technologies, serving as a critical bottleneck. Despite regular updates, these ground systems remain susceptible to cyberattacks. The imperative to adopt a proactive and inclusive approach to cybersecurity is emphasised by the escalating prevalence of cyberattacks directed at space assets. This methodology seeks to provide a strong security framework for satellite ground structures by incorporating widely recognised industry best practices and standards.

However, it should be noted that compliance does not necessarily imply a state of being secure. In the preceding decade, a series of assaults against space ground systems has transpired, accompanied by a deliberate endeavour by adversaries to use internet operations to infiltrate American space technology. The identification and mitigation of risk in satellite ground systems will continue to be significant issues, necessitating the advancement of improved methodologies. Although space ground and control systems may exceed cybersecurity compliance standards, they currently lack a continuous and iterative mechanism for assessing cybersecurity. This deficiency hinders the identification, analysis, and mitigation of advanced cyberspace threats throughout the systems' operational lifespan. To enhance the ground and control systems for space, there should be novel methodologies for risk-based compliance, conduct periodic evaluations of cybersecurity risks associated with space operations, and prioritise the elimination of system faults throughout the design phase.

4.3 Potential Framework Concepts

Since there is very little starting point regarding space governance and previous requirements being driven from the Treaties, there needs to be a structured approach. Technology has advanced at an exponential rate and has multiple dimensions that all carry their risks. These are further complicated by the accelerated advancement of AI and quantum computing and further complicated by the recent surge and huge volume of commercial launching of satellites. There has never been a more required time for a governance framework to be discussed and a plan to implement, monitor and ensure end-to-end security is there.

There will be the usual required steps to implement a governance framework, starting with steering committees, and forums and containing a mix of academics, researchers, government officials, commercial stakeholders, etc. It is possible to take the best qualities from other seasoned frameworks and create and adopt a hybrid framework that is suitable for covering everything from ground to satellite communications, launching satellites, maintaining, cybersecurity and so on.

The following are popular methodologies that can draw on the best and most suitable aspects to start a new framework discussion.

- **COBIT (Control Objectives for Information and Related Technologies**): is a framework developed by the Information Systems Audit and Control Association (ISACA), and provides guidance for effective governance and management of information technology. The framework encompasses four key domains:

    Plan and Organise
    Acquire and Implement,
    Deliver and Support, and
    Monitor and Evaluate.

These domains play a vital role in ensuring the provision of security.

- **CVSS (Common Vulnerability Scoring System):** CVSS entails a scoring system for comparative assessment of the severity of potential vulnerabilities pertaining to cyber security. A 'modified base metrics' is employed to rank the potential vulnerabilities based on the mathematical approximation of the values attached to the respective severity. However, like CMMI, CVSS provides little guidance regarding disaster and recovery planning (CVSS, 2017).

- **The ISO (International Organisation for Standardisation)**: According to ISO 27001, it is imperative to establish, implement, maintain, and enhance an Information Security Management System (ISMS). This promotes a range of international standards for managing recovery from cyber-attacks. For example:

  ISO 27001 provides guidelines on how to establish an Information Security Management System (ISMS);

  ISO 27031 facilitates the development and implementation of recommendations for cyber-attacks related disaster recovery (ISO, 2017);

  ISO 27032 enables the development of a collaborative approach to between supply chain participants to tackle cyber-attacks in a coordinated manner.

  The **Statement of Applicability (SoA)** is an important basis for the treatment of the risks and direction of the mandatory and optional controls to deploy and covers the 10 clauses in ISO/IEC 27001:2013. It provides clear guidance on how to manage the risks and vulnerabilities and forms a basis of the monitoring and auditing process. A particularly important benchmark that all organisations in the supply chain can follow and adopt. It also means any new parties wanting to join the supply chain would need to have some level of understanding of these practises plus be willing to undergo constant monitoring and evaluation of a cycle of continuous improvement.

- **NIST (National Institute of Standards and Technology's Cyber Security Framework):** Under NIST, cyber security activities are divided into 5 distinct categories (e.g., identify, protect, detect, respond, and recover). NIST is the most advanced framework for recovering services compromised via cyber-attacks by providing guidance related to recovery planning, improvements and communications (NIST, 2014).
- **The OCTAVE (Operationally Critical Threat, Asset, and Vulnerability Evaluation):** Informed by qualitative data, this framework offers guidelines for understanding and exploring the potential cyber-risks through a series of workshops (Caralli et al., 2007). Because of its departure from a quantitative approach, OCTAVE is unable to provide an objective measure of the 3 levels of recovery (high, medium, low) required from cyber-attacks.
- **TARA (Threat Assessment & Remediation Analysis):** is a qualitative analytical model that applies threat matrix and standardised template to record system threats from cyber-attacks. TARA promotes and somewhat facilitates the identification of appropriate recovery options, but fails to quantify the impact of cyber risks, which is crucial for deciding on appropriate recovery planning (Wynn et al., 2011).
- **The FAIR (Factor Analysis of Information Risk):** It enables the quantitative measurement and analysis of information risk in financial terms by applying the learning from existing quantitative cyber risk assessment models (e.g., Cyber VaR, RiskLens) (Fair, 2017).
- **The CMMI (Capability Maturity Model Integrated):** This model combines the 5-levels (i.e., initial; managed; defined; quantitatively managed; optimizing) of Capability Maturity Model (CMM). The main limitation of CMMI is that it does not provide any guidance regarding cyber-attack-related disaster management or post-disaster recovery planning (CMMI, 2017).

5 Conclusions

The history of space is a recent one that for many years was mostly driven by a limited number of country powers and Treaties whose focus was mainly of a protective stance related to country security, military and other similar defence topics. With the advancement of technology, over the last decade, and with the desires of commercial companies to accelerate the dominance of satellite coverage, it has also accelerated the risk to security and widened the range of landscape vulnerabilities. Disrupting the satellite domain poses huge risks to individuals, organisations and countries if the end-to-end security is not fully secure, constantly monitored and checks in place to evaluate other associated risks from AI and quantum computing. These risks are no different to what is experienced in normal cloud computing and technology; similar potential bad actors, accidental data leakage, malicious behaviours, rogue states, etc. The added complications are the domain itself of controlling satellites and their data from ground-based to space, malicious crashes, space debris and generally the inexperience of problem scenarios and how to handle and mitigate them. Plus, a scarcity of available threat modelling data and a 'playbook' for this domain.

The factors above are further pronounced as there is not a recent governance framework that has been discussed and validated and at least can be followed as best guidance. Whilst implementing a governance framework will not mitigate all eventualities, it does provide protective layers and also a method to evaluate when circumstances cause things to go wrong. The time has come for a serious discussion on reviewing a fit-for-purpose governance framework that suits the space domain, covering everything from ground to space, factoring in vulnerabilities and how to mitigate them.

References

Bichler, S. F. (2015). *Mitigating Cyber Security Risk in Satellite Ground Systems*. Air Command and Staff College.

Bryce Space and Technology: State of the Satellite Industry Report 2017. Resreport. (2017). https://brycetech.com/downloads/SIA_SSIR_2017.pdf .

Caralli, R. A., Stevens, J. F., Young, L. R., & Wilson, W. R. (2007). *Introducing OCTAVE Allegro: Improving the Information Security Risk Assessment Process*. Hansom AFB.

CCSDS. (2015). *Security Threats Against Space Missions: Report Concerning Space Data System Standards*. https://public.ccsds.org/Pubs/350x1g2.pd.

Chachra, A., & Sharma, D. (2019). *Applications of Machine Learning Algorithms for Countermeasures to Cyber Attacks*. [Online] papers.ssrn.com. Available at https://papers.ssrn.com/sol3/papers.cfm?abstract_id=3370181.

CMMI. (2017). *What Is Capability Maturity Model Integration (CMMI) ®?* CMMI Institute [Online]. Available at http://cmmiinstitute.com/capability-maturity-modelintegration. Accessed November, 2023.

Curzi, G., Modenini, D., & Tortora, P. (2020). Large Constellations of Small Satellites: A Survey of Near Future Challenges and Missions. *Aerospace, 7*(9), 133.

CVSS. (2017). *Common Vulnerability Scoring System SIG.* FIRST.org [Online]. Available at https://www.first.org/cvss/. Accessed November, 2023.

Falco, G. (2018). *Job One for Space Force: Space Asset Cybersecurity.* [Online] Available at https://www.belfercenter.org/sites/default/files/files/publication/CSP%20Falco%20Space%20Asset%20-%20FINAL.pdf.

FAIR. (2017). *Quantitative Information Risk Management: The FAIR Institute.* Factor Analysis of Information Risk [Online]. Available at http://www.fairinstitute.org/. Accessed November, 2023.

Fritz, J. (2013). Satellite Hacking: A Guide for the Perplexed. *Culture Mandala, 10*(1), 5906.

Geospatial World (2023). *How Many Satellites Are orbiting the Earth in 2022.* https://www.geospatialworld.net/prime/business-and-industry-trends/how-many-satellites-orbiting-earth/

Hashesh, A., Hashima, S., Zaki, R., Fouda, M., Hatano, K., & Eldien, A. (2022). AI-Enabled UAV Communications: Challenges and Future Directions. *IEEE Access, 10*, 92048–92066. https://doi.org/10.1109/ACCESS.2022.3202956

Hassija V., et al. (2021). Fast, Reliable, and Secure Drone Communication: A Comprehensive Survey. *IEEE Communications Surveys & Tutorials, 23*(4), 2802–2832, Fourth quarter. https://doi.org/10.1109/COMST.2021.3097916

Hosseinidehaj, N., Babar, Z., Malaney, R., Ng, S. X., & Hanzo, L. (2019). Satellite-Based Continuous- Variable Quantum Communications: State-of-the-Art and a Predictive Outlook. *IEEE Communications Surveys & Tutorials, 21*(1), 881–919. [Online] https://doi.org/10.1109/COMST.2018.2864557 .

Howell, E., & Pultrova, T. (2023). *Starlink Satellites: Everything You Need to Know About the Controversial Internet Mega Constellation.* Space.com. [Online] www.space.com/spacex-starlink-satellites.html

Ibarra, J., Butt, U. J., Do, A., Jahankhani, H., & Jamal, A. (2019, January). Ransomware Impact to SCADA Systems and Its Scope to Critical Infrastructure. In *2019 IEEE 12th International Conference on Global Security, Safety and Sustainability (ICGS3)* (pp. 1–12). IEEE.

ISO International Organization for Standardization. (2017). [Online]. Available at https://www.iso.org/news/ref2451.html. Accessed November, 2023.

Kaul, S. (2017). Need for Space Governance for India and Global Space Governance. *The Journal of Space Safety Engineering, 4*(1), 51–56.

Kodheli, O., Lagunas, E., Maturo, N., Sharma, S. K., Shankar, B., Montoya, J. F. M., Duncan, J. C. M., Spano, D., Chatzinotas, S., Kisseleff, S., Querol, J., Lei, L., Vu, T. X., & Goussetis, G. (2021). Satellite Communications in the New Space Era: A Survey and Future Challenges. *IEEE Communications Surveys & Tutorials, 23*(1), 70–109. https://doi.org/10.1109/comst.2020.3028247

Li, Y., & Liu, Q. (2021). A Comprehensive Review Study of Cyber-Attacks and Cyber Security; Emerging Trends and Recent Developments. *Energy Reports, 7*(7), 8176–8186. https://doi.org/10.1016/j.egyr.2021.08.126

Livingstone, D., & Lewis, P. (2016). *Space, the Final Frontier for Cybersecurity?* [Online] https://www.chathamhouse.org/sites/default/files/publications/research/2016-09-22-space-final-frontier-cybersecurity-livingstone-lewis.pdf. Accessed November, 2023.

Manesh, M. R., & Kaabouch, N. (2019). Cyber Attacks on Unmanned Aerial System Networks: Detection, Countermeasure, and Future Research Directions. *Computers & Security.* https://doi.org/10.1016/j.cose.2019.05.003

Manulis, M., Bridges, C. P., Harrison, R., Sekar, V., & Davis, A. (2020). Cyber Security in New Space. *International Journal of Information Security.* https://doi.org/10.1007/s10207-020-00503-w

McCarthy et al. (1995). *A Proposal for the Dartmouth Summer Research Project on Artificial Intelligence*. [Online]. Available at https://www-formal.stanford.edu/jmc/history/dartmouth/dartmouth.html

NASA Cybersecurity: An Examination of the Agency's Information Security Statement of Paul K. Martin Inspector General National Aeronautics and Space Administration. (2012). Available at http://www.csri.info/wp-content/uploads/2012/08/HHRG-112-SY21-WState-PMartin-20120229-1.pdf.

NATO (2022). *NATO's Overarching Space Policy,* (n.d.). Retrieved July 5, 2023, from https://www.nato.int/cps/en/natohq/official_texts_190862.htm

Newman, C. (2017). The Regulation of Satellites in the 21st Century Space Environment. *Outer Space Law. Global Law and Business*. ISBN 9781911078197

NIST. (2014). *Framework for Improving Critical Infrastructure Cyber Security*. (Online). Available at https://www.nist.gov/publications/framework-improving-critical-infrastructure-cybersecurity-version-11. Accessed November, 2023

Oltrogge, D. L., & Christensen, I. A. (2020). Space Governance in the New Space Era. *Journal of Space Safety Engineering, 7*(3), 432–438.

Pavur, J., & Martinovic, I. (2022). Building a Launchpad for Satellite Cyber-Security Research: Lessons from 60 years of Spaceflight. *Journal of Cybersecurity, 8*(1) [Online] https://doi.org/10.1093/cybsec/tyac008.

Russell, S., & Norvig, P. (2020). *Artificial Intelligence: A Modern Approach* (4th ed.). Pearson Education Limited.

Shakhatreh, et al. (2019). Unmanned Aerial Vehicles (UAVs): A Survey on Civil Applications and Key Research Challenges. *IEEE Access, 7*, 48572–48634. https://doi.org/10.1109/ACCESS.2019.2909530

Sharma, et al. (2020). Communication and Networking Technologies for UAVs: A Survey. *Journal of Network and Computer Applications, 168*, 102739, ISSN 1084–8045. https://doi.org/10.1016/j.jnca.2020.102739.

Sweeting, M. N. (2018). Modern Small Satellites-Changing the Economics of Space. *Proceedings of the IEEE, 106*(3), 343–361.

Tedeschi, P., Sciancalepore, S., & Di Pietro, R. (2022). Satellite-Based Communications Security: A Survey of Threats, Solutions, and Research Challenges. *Computer Networks*, 109246. https://doi.org/10.1016/j.comnet.2022.109246

Tegmark, M. (2017). *Life 3.0: Being Human in the Age of Artificial Intelligence*. Penguin Random House.

Tosun, O. K. (2021). Cyber-Attacks and Stock Market Activity. *International Review of Financial Analysis, 76*, 101795. https://doi.org/10.1016/j.irfa.2021.101795

Thomas, A., & Manuel, E. M. (2016). Embedment of Montgomery Algorithm on Elliptic Curve Cryptography over RSA Public Key Cryptography. *Procedia Technology, 24*, 911–917. https://doi.org/10.1016/j.protcy.2016.05.179

Tsochev, G., Trifonov, R., Nakov, O., Manolov, S., & Pavlova, G. (2020). *Cyber Security: Threats and Challenges*. [Online] IEEE Xplore. https://doi.org/10.1109/ICAI50593.2020.9311369 .

Turner, L. A, & Jahankhani, H., (2021). An Investigation into an Approach to Updating the Governance of Satellite Communications to Enhance Cyber Security. *Cybersecurity, Privacy and Freedom Protection in the Connected World* (pp.23–33). Springer International Publishing

Varadharajan, V., & Suri, N. (2022). *Security Challenges When Space Merges with Cyberspace*. arXiv preprint arXiv:2207.10798.

Wooldridge, M. (2020). *Artificial Intelligence: A Very Short Introduction* (2nd ed.). Oxford University Press.

Wynn, J., Whitmore, G., Upton, L., Spriggs, D., McKinnon, R., McInnes, R., Graubart, L., & Clausen, J. (2011). *Threat Assessment & Remediation Analysis (TARA) Methodology Description Version 1.0*. MITRE.

Yaacoub, J.-P., et al. (2020). Security Analysis of Drones Systems: Attacks, Limitations, and Recommendations. *Internet of Things, 11*, 100218. https://doi.org/10.1016/j.iot.2020.100218

United Nations. Office for Outer Space Affairs & United Nations. General Assembly. (2002). *United Nations Treaties and Principles on Outer Space : Text of Treaties and Principles Governing the Activities of States in the Exploration and Use of Outer Space, Adopted by the United Nations General Assembly.* United Nations.

Defining the Maritime Cyber Adversarial Taxonomy of Tactics and Techniques

Scott James Nash and Hamid Jahankhani

Abstract The maritime industry by design has unknowingly embarked on a digital transformation whereby vessels at sea are an extension of shoreside infrastructure presenting a significant threat of a cyber-attack occurring at sea. The remote nature of the industry has created an undertone of immunity towards a cyber-attack despite common attack vectors being prevalent onboard. The use of specific maritime technologies has created a meshed network whereby a cyber adversary can impact a vessel via multimodal methods of communication, whether this be via satellite communication or ship to ship transmissions. The global nature of the maritime domain is more likely to be impacted by the geopolitical landscape as vessels cross geographic boundaries predominately unhindered, this factor is unique to the maritime industry and should be considered during the development of a maritime threat model. The mapping of adversarial behaviours against the MITRE ATT&CK® taxonomy present a unique perspective of the maritime cyber adversary and the resultant data presents a tangible alignment and overlap with other industries. The outcome demonstrates that the maritime industry is not unique when faced with a cyber threat actor and can draw on similarities to documented taxonomies from the cyber community.

Keywords Maritime · Cyber Security · MITRE ATT&CK · Adversarial Behaviours · Taxonomy

1 Introduction

The maritime domain is one of the oldest forms of transportation which underpins the modern economy. The industry had witnessed a rise in cyber-attacks in recent years which has the potential to cause a significant impact (Meland et al., 2021). A recent survey of 800 maritime professionals concluded that 90% are expecting disruption of a vessel or fleet in relation to a cyber-attack, with the impact is likely to cause a collision, grounding, or environmental damage (DNV, 2023). Despite the

S. J. Nash · H. Jahankhani (✉)
Northumbria University London, London, UK
e-mail: Hamid.jahankhani@northumbnria.ac.uk

H. Jahankhani et al. (eds.), *Space Governance*, Space Law and Policy,
https://doi.org/10.1007/978-3-031-62228-1_2

development of industry wide standards to increase cyber security awareness, they focus on the articulation of risk without clearly identifying the cyber-attack surface or use common adversarial modelling techniques to conduct a threat informed defensive strategy (Caprolu et al., 2020).

The maritime industry has by design and regulation implemented a converged Information Technology (IT) and Operational Technology (OT) network to support the interconnectivity of a sensor network to provide an Integrated Bridge System (IBS). This was conceptualised prior to 1996 and has the objective of increasing situational awareness and safety onboard (International Maritime Organization, 1996). During the development of these systems the basic cyber security principles by modern standard were not considered. The rapid advancements of the internet and Industry 4.0 requires internet connectivity to previously air-gapped systems for telemetry and metric reporting. The maritime industry is undertaking a similar journey of digitisation known as shipping 4.0, combined with the concept of cyber enabled shipping and the creation of maritime autonomous platforms has increased the attack surface (Kavallieratos et al., 2020). The connection of maritime platforms to satellite based communication platforms was a keystone activity and recent events have demonstrated cyber adversary using this mode of communication to deliver traditional IT attack vectors to remote platforms (USCG, 2019). The inadequacies of cyber security onboard maritime platforms have not been addressed since conceptualisation, this is due to the lack of attention by shipping companies and academia combined with a widespread opinion of immunity toward a cyber-attack affecting a maritime platform (Caprolu et al., 2020). The maritime standard and guidelines focus on risk articulation, but do not address the multi-dimensional approach adopted by an adversary targeting the maritime industry (BIMCO, n.d.). This presents a gap whereby the industry is attempting to close the cyber-attack surface using the regulatory landscape without considering the need to adopt a threat informed approach to deliver tangible mitigations based on adversarial mapping.

The maritime industry is perceived to face unique cyber adversaries and threats due to the remote nature, bespoke technologies used, and limited connectivity of shipping platforms. The researcher predicts that mapping of cyber adversarial tactics and techniques will demonstrate a tangible relationship and similarities of other industries.

2 Literature Review

The maritime industry is largely associated with the transportation of commodities by sea, which equates for 90% of world trade (Caprolu et al., 2020). This operation requires multi-organisational infrastructure to support a complex logistics network across a diverse portfolio of industries both at sea and ashore. Cyber Enabled Ships (C-ES) describes the interconnected and converged nature of Information Technology (IT) and Operational Technology (OT) systems which support safety critical systems onboard. Remote connectivity and monitoring by shoreside facilities is enabled by satellite communications, combined with development of maritime autonomous surface ships (MASS) and Autonomous Underwater Vehicles (UAV). These are remotely operated from a Shore Control Centre (SCC) which creates a shared ecosystem, increasing the attack surface which is accessible from anywhere within cyberspace. Similarities can be drawn from Industry 4.0 as the increased use of IoT devices in support of basic functions is diversifying the technology used within the industry creating the concept or term of "shipping 4.0" and "hyper-connected ports" (Darktrace, n.d.). The intermodal digitisation of the maritime industry and network convergence creates a diverse and increased attack surface (Ashraf et al., 2023), which from a cyber security perspective is rarely considered in its entirety or from an adversarial perspective. This is combined with the widespread belief of immunity towards a cyber-attack being held by the maritime community setting a dangerous precedent from a cyber security perspective (Caprolu et al., 2020; Kavallieratos et al., 2020; Weaver et al., 2022).

The industry is a high value lucrative target for cyber criminals with the ocean economy being valued at $3 trillion USD and is anticipated to provide employment for 40million people by 2030 (OECD, 2016). The ramifications of a cyber-attack against essential onboard safety critical systems could endanger human life and/or the marine environment. The socio-economic impact could be much greater, Weaver estimates the financial impact of a cyber-attack against Port Everglade could cost between $80,000 and $2.8million per week in lost revenue dependant on the schedule (Weaver et al., 2022). Computer networks control the critical infrastructure to support the basic life support onboard such as power generation, water desalination and ventilation, drawing similarities to critical national infrastructure. This is defined as "services whose integrity, if compromised, could result in significant loss of life or casualties - taking into account significant economic or social impacts" (NPSA, 2023; Schinas & Metzger, 2023).

2.1 *Onboard Cyber Physical Systems*

The increased reliance on computerised technology to provide essential services such as navigation, communications, passenger, and cargo management have increased the attack surface of vessels provides a visual aid of the complexity and interconnectivity onboard a maritime vessel.

The use of individual components such as Automated Information System (AIS), Electronic Chart Display and Information System (ECDIS), Global Navigation Safety Systems (GNSS) and Radar provide the critical navigational safety functionality onboard (Caprolu et al., 2020). The Integrated Bridge System (IBS) is the central point in an interconnected sensor network and was introduced under MSC 64 (67) in 1996. The aim was to provide "safe and efficient" ship management which integrated two or more operations such as machinery control, safety, and cargo control systems from a dedicated workstation (International Maritime Organization, 1996). The regulation did not consider the cyber security implications of the IBS and was ratified seven years prior to the first cyber-attack on the UK Government (NCSC, 2023). This demonstrates that the maritime industry has been operating a converged network, fundamentally by design and regulation for a significant period. This presents a systemic cyber security risk upon the introduction of shipping 4.0 requirements for internet connectivity which is believed to be provided predominantly by satellite communication. However, Maritime platforms require multiple terrestrial communications methods such as the 802.16e mobile WMAX microwave technology (Silverajan et al., 2018) and the expansion of the Very High Frequency (VHF) radio spectrum for Machine to Machine (M2M) IoT communications. This provides greater coverage than an LTE network and provides a connectivity for low data rate IoT devices to send telemetry, potentially from fixed maritime platforms. Having multiple connectivity mediums to meet the need of continuous connectivity is increasing the cyber-attack surface. (International Maritime Organization, 1996; NCSC, 2023; OFCOM, 2016).

The ability to compromise the cyber physical nature of a maritime vessel's converged flat network (USCG, 2019) would permit an attacker to control crucial peripheral systems causing significant disruption as the ships network is wired for reliability not security (Caprolu et al., 2020; Silverajan et al., 2018). The subsystem sensor network transmits time sensitive information, which is displayed visually on the IBS, however these systems are rarely logically separated, and, in most cases, they transmit data in clear text (Caprolu et al., 2020). This presents a confidentiality and integrity risk, allowing an attacker to intercept, and tamper with the data in transit, making these systems the most vulnerable onboard (Ashraf et al., 2023; Kavallieratos et al., 2020; Wolsing et al., n.d.). Despite recent attacks ship owners and operators are unclear of the obsolete protocols and technologies when compared with modern cyber security requirements and the vulnerabilities and threats which affect modern vessels. (Caprolu et al., 2020).

2.2 Electronic Chart Display and Information System (ECDIS)

The introduction of Marine Information Notice (MIN) 445 released in October 2012 amended the Safety of Life at Sea (SOLAS) regulation permitting the use of EDIS as the primary and backup means of navigation and permitted, but did not address, the basic cyber security controls to protect confidentiality, integrity, and availability. (Helmick, 2020; Maritime & Coastguard Agency, 2013). In 2011, the European Network and Information Security Agency (ENISA) accessed the awareness of cyber security within the maritime industry as "low to non-existent". Caprolu discusses a widely held perception within the industry that cyber adversaries would find it technically difficult to launch a cyber-attack against a maritime vessel (Caprolu et al., 2020). This presents a conflict as regulation has enforced the use of technology without the prerequisite cyber security knowledge, awareness, and training.

ECDIS is a complex system of systems with a multifunctional display which requires three inputs to be effective: position, speed, and compass heading of the vessel (Fig. 1). These are provided by the Global Navigation Safety System (GNSS), compass, and log (speed) (Bhattacharjee Shilavadra, 2021). This creates a physically interconnected sensor network which is overlayed on a nautical navigation chart for situational awareness. The technology used must be compatible with the International Electrotechnical Commission (IEC) and specifically, the maritime navigation and radiocommunication equipment and systems digital interface standard (61,162–1(2016)). The data between devices is transmitted in a serial asynchronous method, creating a single talker multi listener network with each message having the capacity of sending 8data bits without parity (Fig. 1). The protocol is easily identifiable during network sniffing, a technique employed by an adversary during discovery, as data bit 7 (D7) is always transmitted as a 0. The "security password sentence" provides the capability of authenticating a sender prior to the sending of a protected message by embedding a password string. Unfortunately, this message is sent in plain text which can be intercepted and used for an impersonation or replay attack compromising confidentiality and integrity of the network allowing an attacker to send nefarious data which has the potential to compromise safety. (British Standards Institute, 2020; Caprolu et al., 2020; MITRE ATT&CK, n.d.).

Figure 1 demonstrates the basic construct of ECDIS and the underlying components which support its operation. ECDIS platforms require hardware and firmware for the initial boot sequence, compromising this layer will fundamentally compromise the integrity of the entire system (Regenscheid, 2018). The manufacturer's responsibility is covered in IEC61174 and IEC60945, which are the international standards regarding ECDIS performance and testing requirements. These contains little to no information regarding the updating of the hardware, firmware, or operating system, with the focus on ensuring the application and navigational charts remain up to date. IEC61174 is extremely limited for ECDIS performance testing, the observation of version number and manuals is the sufficient for ensuring cyber security. IEC60945 states that software should be audited by a "competent authority" but does not stipulate this testing must be security focused or align to industrial frameworks such as NIST 800–193 platform resilience guidance (Regenscheid, 2018). This has led to the operating system layer and below becoming neglected as shown in Svilicic research which highlighted the use of obsolete operating systems containing significant vulnerabilities relating to SMBv1, providing an attacker the ability to launch malware on the base operating system (Ashraf et al., 2023; British Standards Institute, 2003, 2015; Svilicic et al., 2019).

The issue is further compounded as SOLAS states that the primary ECDIS may run additional software which is critical to bridge operations. However, the secondary or backup ECDIS may run non-critical software such as email applications, provided that the action is approved by the master (Maritime & Coastguard Agency, 2020). This regulatory decision is likely to have significant consequences from a cyber perspective, which could introduce vulnerabilities at the operating system and application layer. This demonstrates that ECDIS is not an isolated lockdown application and can draw similarities to a traditional computing device, increasing the attack surface. A

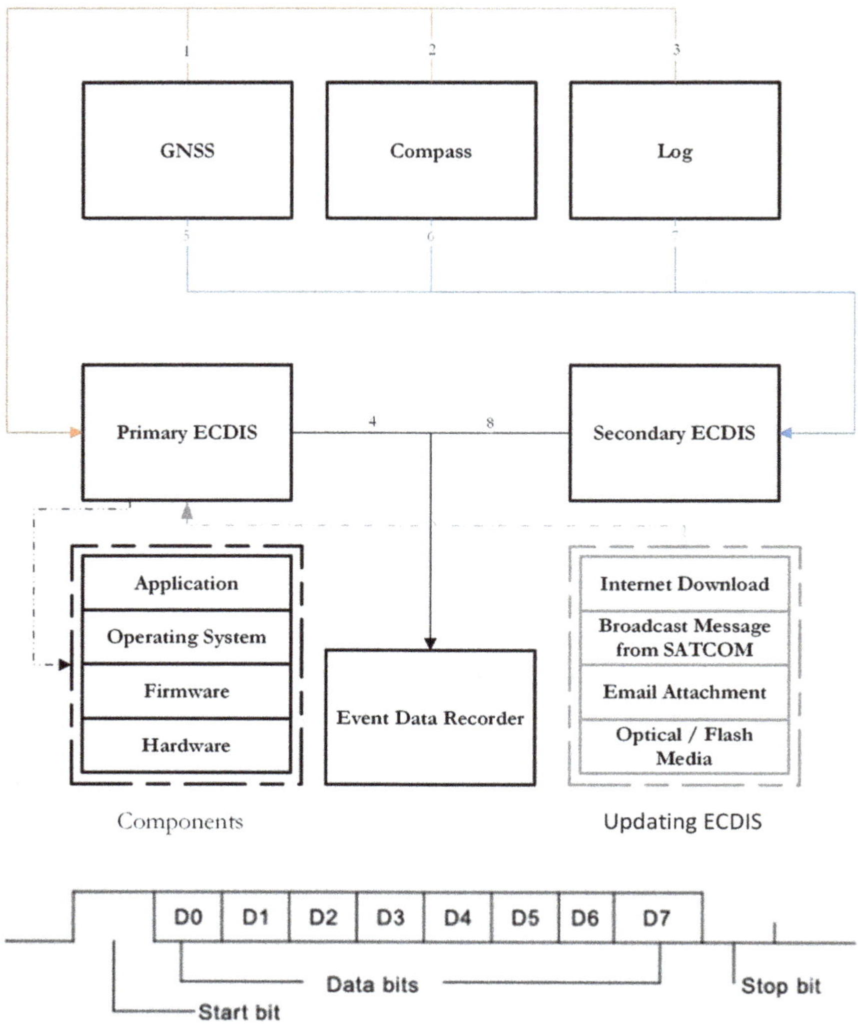

Fig. 1 A logical ECDIS network with a diagrammatic data frame

compromise of the sub sensor network or injection of nefarious data could impact both primary and secondary ECDIS due to the interconnectivity, the standards do not pertain to a physically separate sub system assembly and require the backup to be ready for immediate use (Maritime & Coastguard Agency, 2020). Updating ECDIS presents some significant security concerns as basic adversarial methodologies can be used to establish initial access and persistence whilst moving laterally though the network. The use of phishing emails with attachments containing malware which are transferred to removable media provides a mechanism to compromise the backup

ECDIS, whilst maintaining persistent access via the continuous connection to the satellite communications system (MITRE ATT&CK,). Kavallieratos uses a quantitative and qualitative risk analysis framework based on the DREAD and STRIDE methodology to determine impact and likelihood of a cyber-attack against maritime platforms. The outcome for ECDIS was High impact meaning significant financial damage, or physical damage to infrastructure or loss of human life. The likelihood was high due to the existence of capable threat actors, limited controls, wide variety of vulnerabilities and high exposure to internet connectivity but the regulatory position remains low when addressing cyber security (Kavallieratos & Katsikas, 2020).

2.3 *Event Data Recorder*

The event data recorder or voyage data recorder (VDR) is a regulatory requirement for the storage of sequential records in relation to navigational equipment, position data and sensor information throughout the ships voyage. The predominant literature regarding the VDR focuses on the insider risk as it assumes the attacker must be onboard and connected to the Local Area Network (LAN)) or the use of optical media to inject launch the cyber-attack. The VDR must be connected to the internal network to receive information from the sensor network providing the adversary the opportunity to laterally move from other compromised devices. There is little discussion of defensive capabilities being used within the architecture such as data diodes to provide unidirectional data flow mitigating this vector (Stouffer et al., 2015). The VDR must also be connected to an external time source to mitigate time tampering of the log information, this connection provides an external vector for cyber adversaries to compromise the device, manipulate or erase data. These devices are known to use weak encryption protocols, contain software vulnerabilities and use obsolete firmware which provide the adversary with the mechanisms to launch the cyber-attack ranking this equipment as medium impact when using Kavallieratos methodology (Kavallieratos & Katsikas, 2020; Silverajan et al., 2018; Tam and Jones, n.d.).

2.4 *Automatic Identification System (AIS)*

AIS is used to broadcast the vessels position, speed, and direction to other vessels in the local vicinity (circa 74km), this is conducted using VHF frequencies in a ship-to-ship transmission creating the representation of a localised radio network which can be used by a cyber adversary. Developed in the 1980s as a safety system for collision avoidance, its scope has increased to include many cyber-physical operations such as search and rescue, as well as the monitoring of maritime critical infrastructure. The development of space-based AIS which communicates via Low Earth Orbit (LEO) satellites has increased the range to circa 400km. AIS is described as the most

Table 1 GNSS providers

Country	System
Chinese	BEIDOU
European	Galileo
Russian	GLONASS
United States of America	Global Positioning System (GPS)

vulnerable system onboard; the transmission protocol can be intercepted, tampered with to impersonate, or falsify information which can endanger the safety of vessels within the vicinity. As discussed in Sect. 2.2 AIS uses the same message types transmitting password in clear text. Denial of service attacks can be conducted by flooding the ship-ship network with data to impact the availability of surrounding systems (Ashraf et al., 2023; Caprolu et al., 2020).

2.5 *Global Navigation Safety System (GNSS)*

The Global Navigation Safety System (GNSS) is the second most vulnerable system onboard and is determined as high impact and likelihood of a cyber-attack (Kavallieratos & Katsikas, 2020). The system provides real time location information from Medium Earth Orbit (MEO) satellites which are pivotal for ECDIS to display the position of the vessel in relation to the navigational charts. A secondary use is to provide timing to safety critical systems, the example provided in Sect. 2.3 required the VDR to receive data from an external time source. The message transmission is in clear text and requires no authentication, from a cyber security perspective this provides negative consequences for confidentiality, integrity, and authentication. The common attack vectors relate to spoofing messages containing nefarious information. Signal jamming transmits a higher-powered radio wave on the same frequency as GNSS increasing the signal to noise ratio preventing ships in close proximity to the jamming devices being unable to receive GNSS, many countries have developed their own GNSS (Table 1) this relates to concerns of GNSS being a sovereign capability (Ashraf et al., 2023; Caprolu et al., 2020).

2.6 *The Future*

The invention of quantum technologies has influenced the maritime industry as Imperial College London is developing a quantum accelerometer, which cools atoms to absolute zero where they demonstrate quantum like properties which can be used to measure the vessels position. This system would counter the well documented attack vectors and limitations previously discussed with the traditional GNSS such as spoofing, jamming, and falsified positions (Clarke, 2023; Dunning, 2023). This ensures that the maritime platforms of the future are self-sufficient when determining its position and direction reducing the reliance on external connectivity and tangentially reducing the cyber-attack surface. Despite the well documented threats regarding the lack of encryption for data in transit onboard a maritime platform, the industry faces an increasing challenge with the threats from quantum computing as industrial standards bodies such as National Marine Electronics Association (NMEA) have yet to demonstrate the adoption of quantum resistant cryptography.

3 Adversarial Behaviours and Modelling Techniques

A cyber adversary is defined as "someone or a group that intends to perform a malicious act against another cyber resources" (Edgar & Manz, 2017). The Baltic and International Maritime Council (BIMCO) released guidelines for cybersecurity onboard maritime vessels which defines the types of cyber threat actors and common vulnerabilities associated with the industry. The guideline is the first appearance of technical cyber countermeasures and adversarial behaviour modelling to aid in the development of onboard defensive strategies such as the cyber kill chain which is used to design defence in depth. The document continues to describe the cyber detection capabilities required onboard to indicate a cyber-attack is underway. These include the detection of suspicious network activity and the monitoring of unauthorised administrative or remote logins, these are known as Indicators of Compromise (IOCs) (BIMCO, n.d.; Trend Micro, n.d.).

The development of IOCs is conducted by a Cyber Threat Intelligence (CTI) team who aim to map an adversary's perceived motivations, objectives, and modus operandi by systematically categorising their behaviour as they progressively move through phases of the attack, these are known as Tactics, Technics and Procedures (TTPs). CTI disseminates intelligence at a strategic level to enable senior members of the organisation to understand the cyber threat landscape in which their company or multiple vessels are operating. The use of operational CTI would be required onboard to support the understanding of likely threat actors to target a specific vessel or maritime platform whilst allowing the onboard team to prioritise defensive capabilities. CTI at a tactical level provides the onboard network administrator with IOCs to enable effective triage of data when determining active attacks from background noise (White, 2019). Enabling the detection and response to these events requires the collection of event data, specifically the authentication event logs and network traffic analysis to meet the BIMCO requirements. This document does not discuss the requirement for a Security Event and Incident Management (SEIM) tool onboard or the requirement for engagement with CTI, therefore the likelihood of achieving detection is limited (BIMCO, n.d.).

The sections below provide an overview of CTI methodologies and frameworks which can be applied to the maritime domain with a focus on ATT&CK as this is the chosen methodology for this research paper.

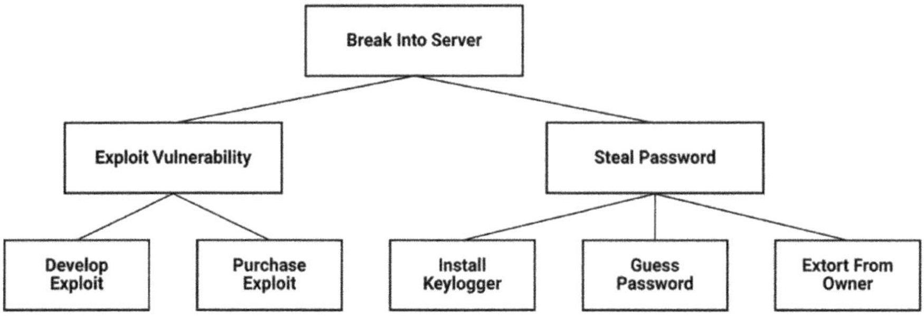

Fig. 2 Attack tree (Stringhini et al., n.d.)

3.1 Attack Trees

Attack Trees aim to visualise a systems security with the root node being the adversaries object and child odes representing techniques they can use to achieve this objective (Fig. 2) (Stringhini et al., n.d.). These models are beneficial when modelling a single system but do not consider observed tactics (IOCs) being used by adversaries, meaning they are predominantly theoretical.

3.2 Lockheed Martin Cyber Kill Chain

The Lockheed Martin Cyber Kill Chain (CKC) is a linear systematic process used to enhance the CTI analyst's understanding of a cyber adversary and is representational of military context (Stringhini et al., n.d.). The focal point is the articulation of Advanced Persistent Threats (APT) and the delivery of malware from outside the network perimeter but is limited when analysing the insider threat which is prevalent concern within the maritime industry. This structured approach demonstrates the end-end process an adversary much complete to achieve a successful cyber-attack. This approach provides defenders the ability to analyse intrusions, extract indicators of compromise and inform the defensive strategy, as preventing an adversary at any stage is perceived to thwart the attack vector and fundamentally stopping the attack. Strategically analysing multiple kill chains over time will identify similarities which can be used to inform frameworks, standards, and policies within the maritime industry (Hutchins et al., n.d.; Shabashnyi, 2015).

3.3 MITRE ATT&CK & Navigator

MITRE Adversarial Tactics, Techniques, and Common Knowledge (ATT&CK) is an in-depth adversarial knowledge base covering IT and Industrial Control System (ICS) domains. The model provides a holistic, technology agnostic observation of a cyber adversary based on empirical data. The ATT&CK matrix has many use cases. The most prevalent for the maritime industry is the ability to conduct defensive gap analysis across IT (Fig. 3) and OT domain (Fig. 4) respectively, whilst using an adversarial approach and common language from the cyber security community enables a threat informed baseline. Tactics are time-agnostic adversarial goals e.g., reconnaissance, whereas techniques are the means of achieving this goal. Sub techniques provide additional information regarding specific means of the technique, including mitigations, these elements are shortened to adversarial TTPs (Pols Paul, 2023; Strom et al., 2020).

Tracking of persistent threat activity is conducted by public and private CTI analysts who evaluate TTPs for similarities to form threat groups, actors or campaigns which are referred to as "Groups" (Fig. 5), understanding how these groups operate within cyberspace is beneficial for the maritime industry as this information can be used to complement the defensive assessments (Strom et al., 2020).

The version of MITRE ATT&CK Enterprise at the time of writing was v13 which had 196 techniques with 411 sub techniques, covering Windows, macOS, Linux, PRE, Azure AD, Office 365, Google Workspace, SaaS, IaaS, Network, and Containers. This demonstrates that ATT&CK contains a broad range and depth of adversarial behaviours which provides a common taxonomy for all industries and is complementary to the non-standard nature of technology deployed in the maritime industry (MITRE ATT&CK, n.d.). Due to the volume of Tactics and Techniques, extracting these from open-source intelligence can be daunting and challenging task. Figure 6, demonstrates the extraction of tactics and techniques which requires a deep understanding of adversarial behaviours and the taxonomy, mistakes can lead to inaccurate categorisations (Applebaum et al., n.d.).

"Decider" is a container-based web application developed by the United States Cybersecurity & Infrastructure Agency (CISA) to aid in the mapping of adversarial behaviours which uses a guided questioning methodology of the ATT&CK taxonomy to mitigate some of the inaccuracy concerns previously raised (CISA, 2023a).

The relationship between tactics, techniques, and sub-techniques can be visualised in ATT&CK Navigator. This tool allows CTI analysts to aggregate, compare and annotate layers which can be used to prioritise prevalent techniques in a tabular form. This process demonstrates a tangible benefit for CTI to identify and analyse threat groups who can pass tangible intelligence to defensive teams to implement detections and mitigations for valid accounts (Applebaum et al., n.d.; MITRE, n.d.). Whilst tabular representation and combined functionality of exporting to data to csv, json and svg formats for future analysis is beneficial ATT&CK is limited to colourisation.

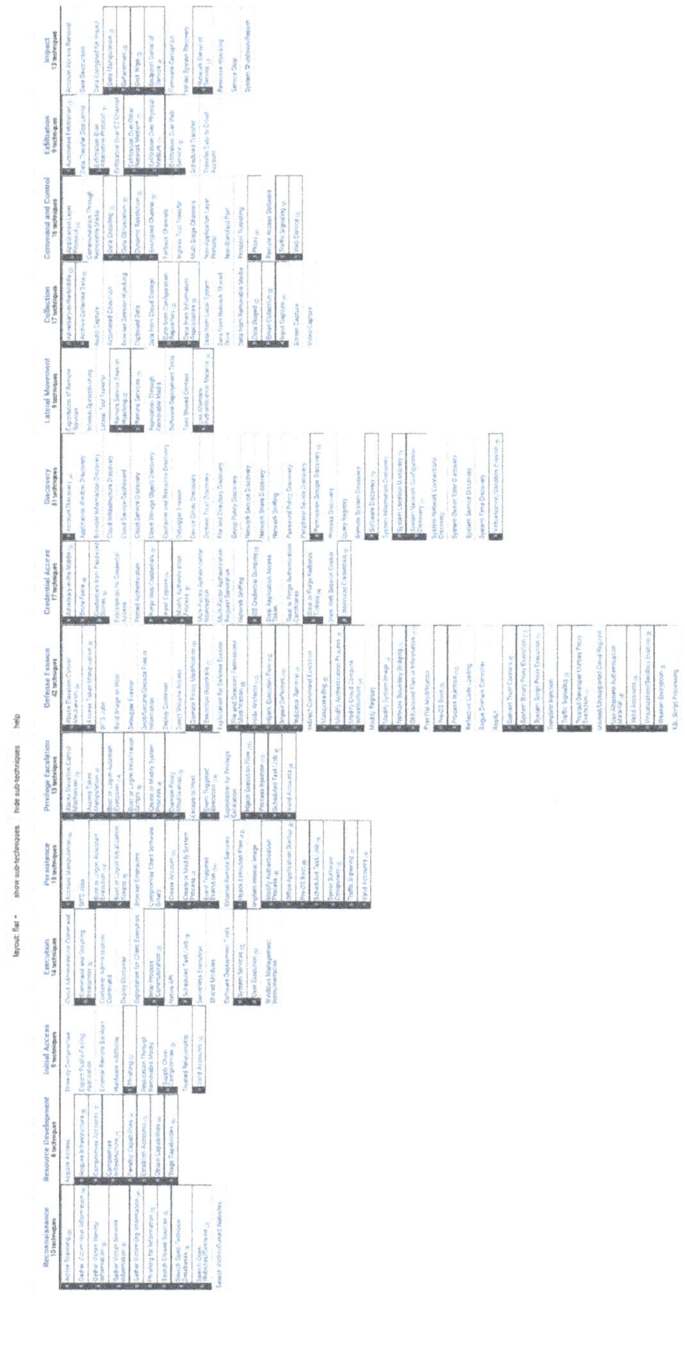

Fig. 3 MITRE ATT&CK enterprise matrix (MITRE ATT&CK, n.d.)

Initial Access 12 techniques	Execution 9 techniques	Persistence 6 techniques	Privilege Escalation 2 techniques	Evasion 6 techniques	Discovery 5 techniques	Lateral Movement 7 techniques	Collection 11 techniques	Command and Control 3 techniques	Inhibit Response Function 14 techniques	Impair Process Control 5 techniques	Impact 12 techniques
Drive-by Compromise	Change Operating Mode	Hardcoded Credentials	Exploitation for Privilege Escalation	Change Operating Mode	Network Connection Enumeration	Default Credentials	Adversary-in-the-Middle	Commonly Used Port	Activate Firmware Update Mode	Brute Force I/O	Damage to Property
Exploit Public-Facing Application	Command-Line Interface	Modify Program	Hooking	Exploitation for Evasion	Network Sniffing	Exploitation of Remote Services	Automated Collection	Connection Proxy	Alarm Suppression	Modify Parameter	Denial of Control
Exploitation of Remote Services	Execution through API	Module Firmware		Indicator Removal on Host	Remote System Discovery	Hardcoded Credentials	Data from Information Repositories	Standard Application Layer Protocol	Block Command Message	Module Firmware	Denial of View
External Remote Services	Graphical User Interface	Project File Infection		Masquerading	Remote System Information Discovery	Lateral Tool Transfer	Data from Local System		Block Reporting Message	Spoof Reporting Message	Loss of Availability
Internet Accessible Device	Hooking	System Firmware		Rootkit	Wireless Sniffing	Program Download	Detect Operating Mode		Block Serial COM	Unauthorized Command Message	Loss of Control
Remote Services	Modify Controller Tasking	Valid Accounts		Spoof Reporting Message		Remote Services	I/O Image		Change Credential		Loss of Productivity and Revenue
Replication Through Removable Media	Native API					Valid Accounts	Monitor Process State		Data Destruction		Loss of Protection
Rogue Master	Scripting						Point & Tag Identification		Denial of Service		Loss of Safety
Spearphishing Attachment	User Execution						Program Upload		Device Restart/Shutdown		Loss of View
Supply Chain Compromise							Screen Capture		Manipulate I/O Image		Manipulation of Control
Transient Cyber Asset							Wireless Sniffing		Modify Alarm Settings		Manipulation of View
Wireless Compromise									Rootkit		Theft of Operational Information
									Service Stop		
									System Firmware		

Fig. 4 MITRE ATT&CK ICS matrix (MITRE ATT&CK, n.d.)

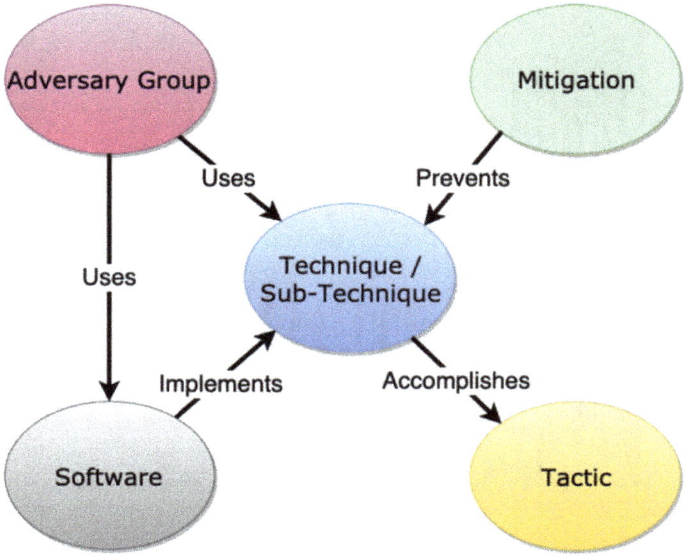

Fig. 5 ATT&ck relationship (Strom et al., 2020)

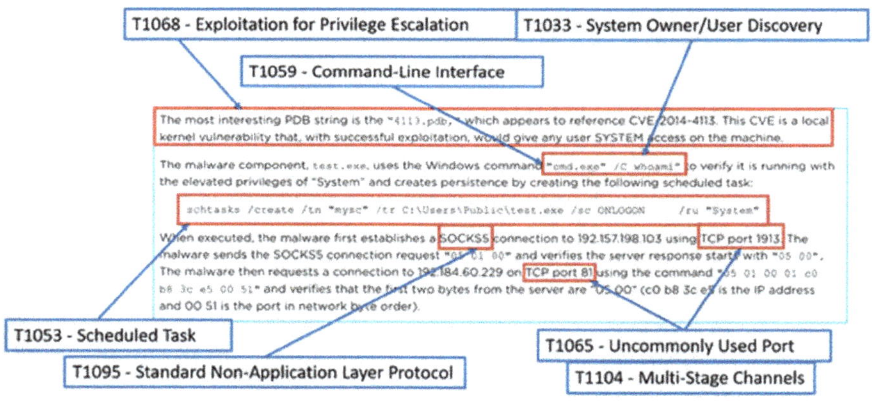

Fig. 6 Report mapping to MITRE ATT&CK (Applebaum et al., n.d.)

3.4 *Unified Kill Chain*

The Unified Kill Chain (UKC) builds on the previous models including the incorporation of the CIA triad, whilst presenting social engineering, exploitation and pivoting as additional parameters. UKC removes the assumption that cyber-attacks must be conducted in a structured manner and the convention of applying mitigations at a single-phase of the kill chain thwarts an attack. UKC presents a perception that tactics used by an adversary may occur out of sequence or be bypassed all together, and instead presents a logical set of distinct grouped categorisations which aid the defence in depth strategy. Challenging the out of sequence assumption ensures that defenders do not focus all their resources on preventing a cyber-attack as early as possible and encourages the mapping of detections and countermeasures to discrete phases to maximises **their effectiveness** (Pols Paul, 2023**).**

4 Data Collection and Critical Discussions

Intelligence analysis is the process of examining data from a wide range of sources with the objective of viewing the complete picture in relation to a situation. The following chapters will use a CTI methodology to enable the direction, collection, processing, and analysis of a product which evaluates the motivations and capabilities of the maritime cyber threat actor's TTPs to enable a threat informed defensive strategy for the maritime industry (Bank of England, 2016; Cabinet Office, 2023; White, 2019).

The threat intelligence lifecycle is a circular process originally developed for military use. The methodology is embedded in civilian sectors such as the UK financial system which enforces the use of the CBEST framework. This was developed by the Bank of England to enable an intelligence led, risk informed approach when protecting the stability of the financial markets. CBEST is the de facto standard for industries and organisation who wish to operationalise the use of CTI. (Bank of England, 2016). The maritime industry has adopted a cyber risk approach but has yet to consider the CTI elements. This thesis aims to demonstrate the use of CTI practices within the maritime industry commencing with phase 1 through to 4 and phase 5 & 6 acting as the submission to the university.

Phase 1—Threat Landscape.

Understanding the cyber threat landscape provides a broad picture of the risks, trends, and weaknesses in relation to the semantics of the industry. 2 provides a threat landscape for the maritime industry, combined with some additional factors below (Bank of England, 2016).

The maritime industry is:

- Relatively small with circa 9800 vessels over 100 gross tonnes which operate internationally (Meland et al., 2021).
- Usually operates vessel for 25–30 years, with major maintenance periods every five years, creating a diverse environment for IT and OT equipment onboard (Meland et al., 2021).
- Regulated internationally which has provided limited focus to technical cyber security, leading to the platforms being vulnerable (Meland et al., 2021).
- Complex from regulatory position as a vessel can be owned by a company in a specific geographic region, registered with a classification society in a different

region and at the time of the cyber incident be in sovereign waters of another nation. This creates complexity understanding which legal justification applies.
- Has had little academic engagement (Direnzo et al., n.d.).

The UK Government Guidance for organisations who are adopting a CTI approach states that a "threat assessment" is required during phase 1 and a mature organisation should aim to:

- Assess only the threat actors with the capability and motivation to attack the organisation.
- Assess specific threat groups based on capability and motivation and not all groups with a label such as "foreign intelligence services (FIS)"
- Provide an in depth understanding of the business assets to enable context of an impact of a cyber-attack (White, 2019).

Due to the limited academic literature and low maturity for CTI methodologies throughout the maritime industry a full threat assessment is not available, this research aims to contribute to raising awareness in the benefits of CTI practices.

Phase 2—Direction.

The direction of CTI aims to establish the requirements early in the lifecycle, this ensures a clear strategy is provisioned for phase 3 (data collection) (Bank of England, 2016; Collier et al., 2023). The requirement for this thesis is to leverage empirical data from academic and external sources to perform historical analysis of cyber-attacks in relation to the maritime domain. An analytic focus will identify ATT&CK Tactics and Techniques used by adversaries to enable trend analysis.

5 Inclusion Criteria

The conditions under which data can be considered suitable for inclusion are detailed below.

- The cyber-attack must:
- have occurred in the last 3 years (January 2020 – October 2023).
- be in relation to and /or impact the maritime industry.
- be a real cyber-attack.
- been successful from the adversary's perspective.

The incident report must:

- contain sufficient information to extract adversarial Tactics and Techniques in relation to the ATT&CK taxonomy.
- be supported by multiple sources for realisability and validity.

The collection must identify:

- a minimum of 20 cyber-attacks be considered acceptable to proceed to the critical analysis. This was agreed in **Error! Reference source not found.**.

6 Exclusion Criteria

The conditions under which data can be considered not suitable and are excluded are detailed below:

- Social Media data is excluded.
- This is due to ethical and privacy concerns whilst adding unnecessary complexity and unreliability.
- Penetration Testing reports are considered out of scope as these are not real cyber-attacks conducted by a motivated adversary.
- A cut-off date was implemented to enable the critical analysis to be undertaken, articles identified after October 2023 were not evaluated.

Phase 3—Primary Sources.

The collection of data was conducted using a chain referral technique, specifically the exponential non-discriminative snowball sampling technique (Meland et al., 2021; "Snowball sampling," n.d.). This requires the identification of initial sources which are used to locate additional relevant data shown in **Error! Reference source not found.**

Three initial sources were selected for the purpose of this research:

- Academic – The Literature discussed above was reviewed for cyber incidents.
- Government – The United States Coast Guard contributes to the Information Sharing and Analysis Centre (ISAC) for the maritime industry. They distribute annual reports, alerts, and guidance. These documents were reviewed for cyber incidents (USCG, n.d.).
- Open-Source—The Maritime Cyber Attack Database (MCAD) was produced by NHL Stenden, University of applied science. They provide an open-source database of maritime cyber security incidents (NHL Stenden University, n.d.).

The initial sources provide a diverse sample of secondary data which includes grey literature. They provide a broad sample with differing perspectives and motivations to obtain cyber incident data. A total of 63 cyber incidents were identified for evaluation across three categories (**Error! Reference source not found.**). This approach aligns with the ATT&CK methodology for OSINT gathering and allows for the extraction of Tactic and Technique to align with the taxonomy (Applebaum et al., n.d.). The limiting of primary sources ensures a tangible outcome is produced within the time constraints. The future work section (**Error! Reference source not found.**) discusses the opportunity to broaden the primary sources using the same technique with the objective of identifying more incidents to enable further quantification of tactics and techniques.

6.1 Collection Methodology

This section aims to demonstrate the process shown in Fig. 7, using a source authored by the US Coast Guard. This guide will demonstrate the collection, initial document review, cataloguing and extraction of ATT&CK Tactics and Techniques from a data source.

The 2022 annual report from the US Coast Guard (United States Coast Guard Cyber Command, 2022) has been identified as a credible source for incident extraction

The collection of adversarial behaviours has analysed 63 initial sources of which 60 were inside the permitted date range. 36 sources provided relevant information for the extraction of 53 unique techniques, achieving the inclusion criteria. A metrics dashboard was maintained throughout to enable consistent mathematical analysis during this and future phases (Fig. 8).

A data frame was constructed to capture ATT&CK Tactics, Techniques, Technique ID, and the observed count. The results contained in Fig. 9 enable the researcher to evaluate the hypotheses and present recommendations in 0. The assimilation of data demonstrates that the impact of a cyber-attack against the maritime industry is easily identifiable. The technique, "data encryption for impact" was observed thirteen times and is likely linked to phishing as the most common initial access vector which was observed six times.

The use of data visualisation enables the conveyance of trends, relationships and the identification of keystone techniques, whilst overlaying the data from multiple extractions. The conveyance of technical information to non-technical audiences is considered a complementary skill of an intelligence analyst. The Professional Head of Intelligence Analysis (PHIA) framework describes the use of data visualisation to communicate with their audiences using of cutting-edge tooling whilst drawing on knowledge and expertise to present a finding (Cabinet Office, 2023; IBM, n.d.; Stringhini et al., n.d.).

These techniques have been implemented to create a visualisation of the data frame using MITRE ATT&CK Navigator ("ATT&CK® Navigator," n.d.). This is an industry tool used for the annotation and trend analysis of adversarial techniques. Navigator can be used manually for small scale adversarial mapping, usually a single incident but as this research aims to evaluate 37 incidents using a single chart, a comma separated JSON file was created to conduct the analysis. The original proposal suggested the creation of the data visualisation using python, due to the data structure and wide variety of tactics identified the use of a network visualisation would be inappropriate, altering the methodology of this objective but does not materially change the outcome.

The final stage of collection resulted in the export of a JSON file which was used to compile the outputs into a single JSON file. Additional parameters are required to render the data visualisation such as the ATT&CK taxonomy version.). The hypothesis aims to evaluate alignment with other industries, therefore the ATT&CK enterprise taxonomy is defined with all platforms selected for presentation.

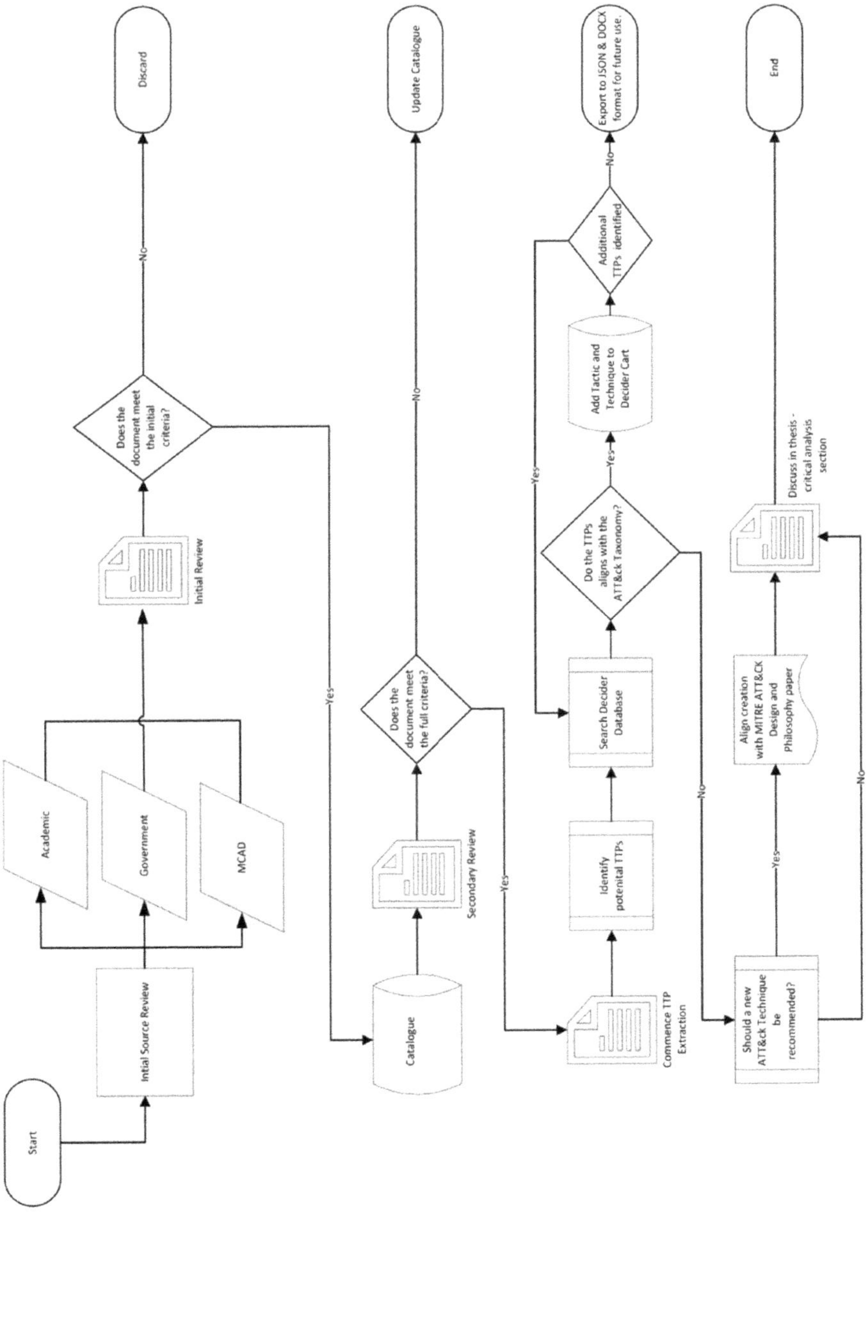

Fig. 7 Data collection and extraction methodology

Data Dashboard

Total number of Incidents
63

Count of Cyber Incidents with successful TTP extraction
36

Count of Cyber Incidents to be discussed in the critical analysis
5

Count of Cyber Incidents Per Year	
Year	Count of Date
2015	1
2017	1
2018	1
2019	1
2020	19
2021	13
2022	12
2023	15
Grand Total	63

Row Labels	Count of Source Type
Academic	18
Government	15
MCAD	30
Grand Total	63

Fig. 8 Metrics dashboard

6.2 Critical Discussion

This section will interpret the results collected from secondary data regarding cyber-attacks against the maritime industry. The evaluation will determine a conclusion in relation to the hypothesis whilst identifying keystone tactics and techniques in relation to the maritime cyber adversary. Due to the constraints of this research this section is limited to 5 of 53 unique techniques and future work has been identified for the continuation and extrapolation of data.

The resultant data has demonstrated that thirty-six of the cyber-attacks reported via open source allowed for the successful extraction of eighty-seven ATT&CK TTPs, this allows the maritime industry to benefit from the detection mechanisms and mitigating controls implemented by other enterprise industries. The subsequent section will use five keystone techniques that will demonstrate that despite the maritime industry being remote in nature and using some bespoke technologies that, the cyber threats and risks posed are tangentially linked to other industries.

- Phishing

The data visualisation clearly shows that a cyber adversary targeting the maritime industry is likely to conduct reconnaissance via email to obtain information. The adversary will gain initial access via a malicious attachment or link contained within a phishing email. These techniques were observed eight times throughout the research making it the most common initial access vector which aligns with the European Union Cyber for Cyber Security (ENSIA) 2022 cyber threat landscape assessment

Fig. 9 Results

Tactic	Technique Name	Technige ID	Count
Reconnaissance			
	Search Victim-Owned Websites	T1594	1
	Gather Victim Identity Information: Credentials	T1589.001	1
	Gather Victim Host Information: Client Configurations	T1592.004	1
	Phishing for Information: Spearphishing Link	T1598.003	1
	Section Total		4
Resource Development			
	Acquire Infrastructure: Domains	T1583.001	2
	Compromise Infrastructure: Domains	T1584.001	1
	Compromise Accounts: Social Media Accounts	T1586.001	1
	Develop Capabilities: Malware	T1587.001	1
	Obtain Capabilities: Vulnerabilities	T1588.006	2
	Stage Capabilities: Upload Malware	T1608.001	1
	Section Total		8
Initial Access			
	Phishing	T1566	2
	Phishing: Spearphishing Attachment	T1566.001	3
	Phishing: Spearphishing Link	T1566.002	1
	Replication Through Removable Media	T1091	2
	External Remote Services	T1133	1
	Drive-by Compromise	T1189	1
	Exploit Public-Facing Application	T1190	1
	Section Total		11
Execution			
	Command and Scripting Interpreter: Windows Command Shell	T1059.003	1
	User Execution: Malicious File	T1204.002	1
	System Services: Service Execution	T1569.002	1
	Section Total		3
Persistence			
	Create or Modify System Process	T1543	1
	Create or Modify System Process: Windows Service	T1543.003	1
	Boot or Logon Autostart Execution: Registry Run Keys / StartupFolder	T1547.001	1
	Section Total		3
Privilege Escalation			
	Section Total		0
Defense Evasion			
	Obfuscated Files or Information	T1027	1
	Hide Artifacts: Run Virtual Instance	T1564.006	1
	Impair Defenses: Disable or Modify Tools	T1562.001	1
	System Binary Proxy Execution: Msiexec	T1218.007	1
	System Binary Proxy Execution: Regsvr32	T1218.010	1
	System Binary Proxy Execution: Rundll32	T1218.011	1
	Section Total		6
Credential Access			
	Brute Force	T1110	1
	Multi-Factor Authentication Request Generation	T1621	1
	Section Total		2
Discovery			
	File and Directory Discovery	T1083	1
	Peripheral Device Discovery	T1120	1
	System Location Discovery	T1614	1
	Section Total		3
Lateral Movement			
	Replication Through Removable Media	T1091	2
	Section Total		2
Collection			
	Data from Local System	T1005	1
	Input Capture: Keylogging	T1056.001	1
	Section Total		2
Command and Control			
	Application Layer Protocol: Web Protocols	T1071.001	1
	Multi-Stage Channels	T1104	1
	Section Total		2
Exfiltration			
	Automated Exfiltration	T1020	1
	Exfiltration Over C2 Channel	T1041	5
	Exfiltration Over Web Service	T1567	1
	Exfiltration Over Web Service: Exfiltration to Cloud Storage	T1567.002	1
	Section Total		8
Impact			
	Data Destruction	T1485	1
	Data Encrypted for Impact	T1486	13
	Service Stop	T1489	1
	Inhibit System Recovery	T1490	1
	Defacement	T1491	1
	Network Denial of Service	T1498	4
	Network Denial of Service: Direct Network Flood	T1498.001	2
	System Shutdown/Reboot	T1529	2
	Data Manipulation: Transmitted Data Manipulation	T1565.002	4
	Data Manipulation: Runtime Data Manipulation	T1565.003	4
	Section Total		33
Total Techniques			87
Unique Techniques			53

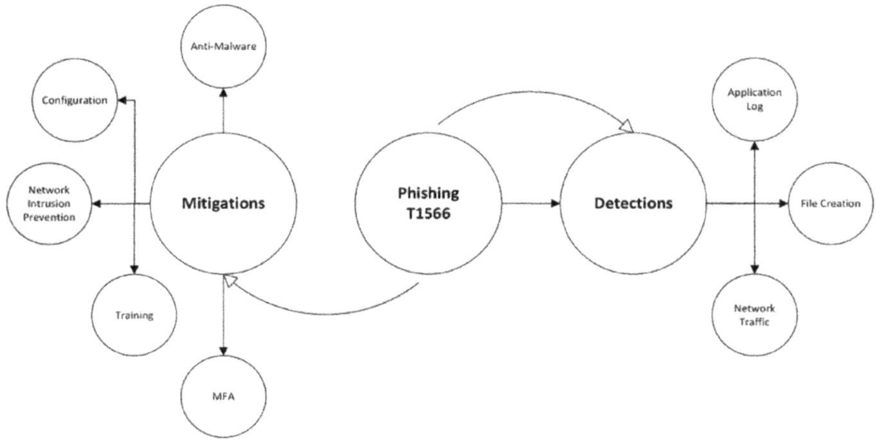

Fig. 10 Phishing Mitigations[1]

(Claudio Ardagna et al., 2022). This demonstrates that the maritime industry faces the same threats as other industries despite the remote nature of its operation.

The navigation system (EDIS) onboard a maritime platform can legitimately be updated via attachments contained within an email (Fig. 1), this presents an opportunity for the adversary to combine social engineering and impersonation to deliver the malicious payload to the victim during the updating period. The maritime guidance for cyber security does clearly articulate the detection opportunities or technical defence to be implemented when attempting to mitigate the threats associated with phishing (BIMCO, n.d.). However due to the alignment with the ATT&CK taxonomy, maritime professionals can easily reference industry best practices in relation to detection and mitigating actions, these are discussed below and shown in Fig. 10

- Mitigation

Defence in depth techniques can be used to provide mitigation for phishing attacks. Configurating the email server to conduct filtering of messages based on validity reduces the likelihood of spoofed emails reaching the users mailbox. This is undertaken by ensuring the senders domain is authentic using Sender Policy Framework (SPF) which provides a list of the senders IP addresses for cross referencing. DomainKeys Identified Mail (DKIM) automatically sign email using public key cryptography enabling authenticity checks to occur when receiving email. Domain-based Message Authentication, Reporting and Conformance (DMARC) sets the policy for the receiving email server and instructions should an email fail SPF or DKIM, an example action would be to quarantine the email preventing it from appearing in

[1] The diagram has been produced by the author with reference to (MITRE ATT&CK, n.d.) and (Australian Cyber Security Centre (ACSC), n.d.).

the user's mailbox. Signature, and heuristic based antimalware software should be configured to scan attachments for malware aiding in the detection of malicious attachments. Additional protections regarding the user's identity should be implemented to reduce the likelihood of identity compromise by configuring multi-factor authentication (Australian Cyber Security Centre (ACSC), n.d.; MITRE ATT&CK, n.d.).

- Detection

Detecting malicious activity can be challenging and requires the monitoring of security audit logs from hardware, software, and applications installed on the platform. The data collected, combined, and retained in a data lake is correlated and analysed to provide context in near real time. The matching of anomalies or Indicator of Compromise (IoC) generated from CTI provide an alerting mechanism for a cyber professional to investigate the potential compromise. The average time to detect a cyber incident is 101 days, therefore it is recommended that security audit data is retained for a minimum of 6 months to enable security investigation to be undertaken. These are commonly referred to as a Security Information and Event Management (SIEM) systems (NCSC, 2018). The maritime guidance relating to detection capabilities is extremely limited, BIMCO references the use of an Intrusion Detection System (IDS) and Intrusion Prevention System (IPS) whilst stating "Logging should be implemented in the firewall and ideally in all network-attached devices" but does not stipulate where the security data should be stored or the minimum retention period (BIMCO, n.d.). Due to the remote nature and intermittent connectivity of the maritime industry, it is recommended that a SIEM is installed, maintained, and monitored onboard.

- Replication through removable media

The use of removable media is prevalent within the maritime industry, with two incidents being identified during this research. Removable media is generally used to update the primary navigation system but also can be used to update OT devices onboard which may be air-gapped for safety reasons. The software updates are provided by the original equipment manufacturer (OEM) or chart updating authority can downloaded via the internet and transferred to removable media. Alternatively, these can be sent directly to the vessel or hand carried by the maintenance engineer. The regulatory requirements regarding VDRs requires a common user interface such as a USB port to be easily accessible in the event accident. These scenarios present the opportunity for malware to replicate through removable media and potentially gaining access to air-gapped systems. This could impact the vessels navigational safety by rendering the systems inoperable, provide an adversary remote access to the vessel or impact the safety of OT devices onboard. Research by Plymouth university has demonstrated that the confidentiality and integrity of data stored on the VDR can be impacted when delivering a malicious payload via USB using a Rubber Ducky (Harish, 2022). This presents a concern from an insider threat perspective which is rarely considered or discussed in the maritime industry as a walled garden approach is untaken using the physical security applied at a port as assumed security. An

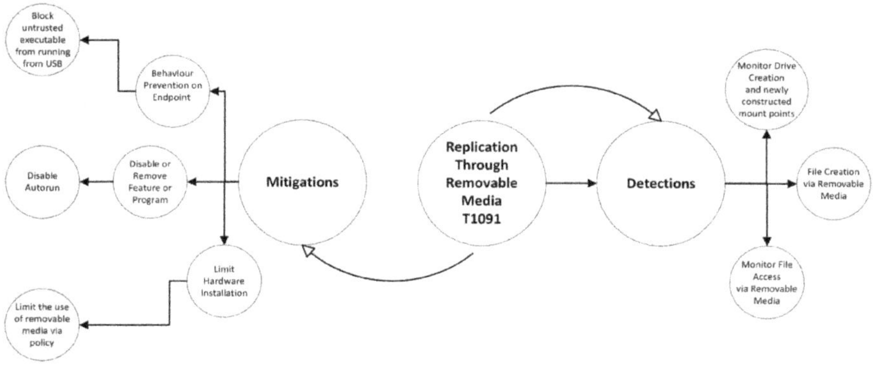

Fig. 11 Replication through removable media mitigation[2]

employee, third party or guest could maliciously tamper with systems onboard and is unlikely to be detected. Should an insider attack be undertaken against the ECDIS this could present a significant safety issue, whereas if conducted against the VDR this could be intentional to destroy evidence.

- Mitigation

Figure demonstrates the procedural and technical controls to be implemented. The limiting of the use of removable media via policy is a procedural control and technically preventing the auto run features and the execution of untrusted executables within the operating system should aid in the prevention of malware execution (MITRE ATT&CK, n.d.). An additional recommendation would be the installation and use of a "sheep-dip" device, a term used within cyber security relating to a terminal which scans removable media using signature and heuristic based anti-malware engines to identify new files which contain malware prior to entering the onboard network. These devices are placed outside the network which can be used by onboard staff and third parties to scan removable media providing additional assurance (OPSWAT, 2018).

- Detection

The detection recommendations shown in Figure relate to the operating systems being configured to monitor the creation of new drive volumes and files which are accessed via removable media (MITRE ATT&CK, n.d.) (Fig. 11).

- Concerns

The detection and monitoring of security logs remains a challenge for maritime industry. However, it relies on suitably qualified and experienced people who are responsible for analysing and identifying a potential cyber-attack from the collected

---

[2] The diagram has been produced by the author with reference to (MITRE ATT&CK, n.d.).

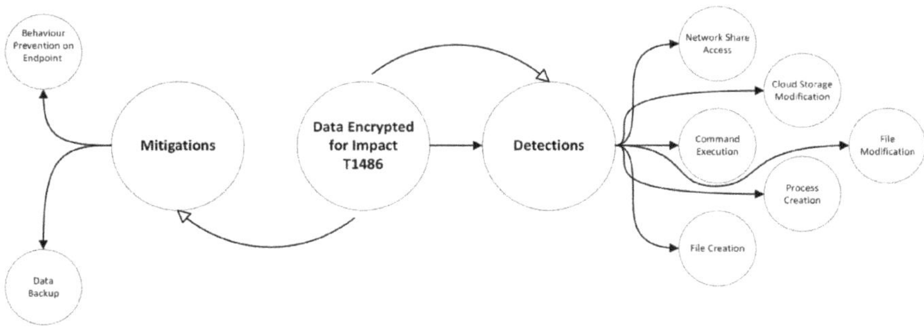

Fig. 12 Data encrypted for impact[3]

logging information. This should be harmonised across the onboard and shoreside infrastructure to ensure full coverage is achieved. Companies operating shoreside in relation to the maritime industry will have a dedicated Chief Information Security Officer who is responsible for developing and leading the cyber security programme, this includes security operations and monitoring teams. These requirements are not translated to the onboard staff, the Electro-Technical Officer (ETO) is required to "operate computers and computer networks onboard" by demonstrating a "basic understanding of local area networks" but no mention of cyber security requirement prior to achieving their certification. The lack of monitoring and competent individuals to identify a cyber-attack onboard presents a reason for the lack of cyber incident reporting at sea.

- Data Encrypted for Impact

Data encryption for impact relates to an adversary impacting the availability of systems in a network by rendering stored data inaccessible, this technique was the most prevalent throughout the research. Ransomware was frequently observed as files were commonly encrypted and victims extorted for financial gain. Ransomware commonly spreads throughout the network via self-replication, using worms like properties (Barrett & Weiss, 2017; MITRE ATT&CK, n.d.). The flat and converged nature of the onboard network presents a significant risk, combined with the regulatory position which permits the use of additional software on critical safety systems such as ECDIS whilst requiring the secondary navigation systems to be available for immediate use. This presents a position whereby ransomware could impact the vessels safety and seaworthiness by encrypting the primary and secondary ECDIS during critical navigation rendering it inoperable for maximum effect. The detection capabilities are similar to the one previously presented demonstrating the benefits of alignment with the ATT&CK taxonomy. The mitigation for ransomware relies on the ability to restore the systems operation from a recent backup. The maritime industry would benefit from adopting the 3–2–1 theory which is commonly presented as three copies of the data, spanning two different media devices and 1 being immutable to

[3] The diagram has been produced by the author with reference to (MITRE ATT&CK, n.d.).

provide adequate ransomware protection and the capability to restore critical systems at sea (Rick Vanover, 2021) (Fig. 12).

- Data Manipulation: Transmitted Data Manipulation

The term spoofing relates to cyber security and is described as the falsification of data to gain an adversarial advantage (CompTIA, n.d.). During the research two services were commonly spoofed, GNSS and AIS. The former impacts situational awareness for the vessels operation as this service is connected to the onboard navigational system. The latter is widely visible to externals and can be freely viewed on the internet. The research identified AIS spoofing predominately occurred against military vessels which can be attributed to geopolitically motivations and disinformation campaigns as military vessels positions were falsified to demonstrate inaccurate positions in relation to disputed territories (Bjorn Bergman, 2021).

GNSS spoofing transmits a falsified signal which has a similar power and structure to the satellite service. This signal is mistaken by the receiver as the correct signal therefore falsifying the position data (Meng et al., 2022). The spoofing of AIS is similar but due to the lack of verification conducted by the transceiver data can be manipulated or falsified. This can have disastrous consequences for the maritime industry as GNSS and AIS are relied upon for navigational safety (Louart et al., 2023; Meng et al., 2022). The research has identified an example whereby a ship was placed 32km away from their original position (Bjorn Bergman, 2021), which creates confusion and a loss of situational awareness which could result in the grounding or collision of the vessel. GNSS is widely adopted by other industries, the security concerns in relation to confidentiality, integrity and lack of authentication are concerns for any industry utilising satellite positioning systems for safety critical acts and therefore is not unique to the maritime industry. The act of spoofing transmitted data, aligns with the MITRE ATT&CK taxonomy as the data is being manipulated in transit for adversarial gain.

- Network Denial of Service

The term Denial of Service (DoS) is defined as the degradation or blocking of a service by exhausting the network bandwidth (MITRE ATT&CK, n.d.). During the research this was observed four times predominantly attacking shore-based infrastructure which was aligned with the geopolitical landscape. Similarities and overlaps with the jamming can be discussed as in the vessels at sea require radio signals carrying critical safety information such as GNSS, preventing these signals from being received would impact the vessels safety and regulatory position in regards of seaworthiness. An example of deliberate signal jamming was identified, the adversary's motivation was to impact the availability of the vessel as it was prevented from leaving harbour without a functioning GNSS (Gregor Gowan, 2022). Jamming is conducted by transmitting using high power on a similar radio frequency with the aim of reducing the signal to noise ratio of the receiver rendering it unable to receive the required signal (Meng et al., 2022). This adversarial act aligns with the MITRE ATT&CK taxonomy as a network is being used to deny the GNSS service, therefore the maritime industry is suspectable to a DoS at sea and ashore demonstrating that despite the unique deployment of technology the adversarial techniques are not unique.

- Undersea Infrastructure

The empirical data reviewed has identified an area which is commonly overlooked in relation to the maritime industry. The undersea communication infrastructure spans 1.3million km and carries 99% of the world digital communications. The most common type of physical damage to this infrastructure is accidental. Deliberately damaging this infrastructure would have disastrous consequences for time sensitive operations which rely on the internet and would cause long lasting economic damage. During the research an article was identified which discusses the potential for deliberate damage of an undersea infrastructure in Norway (ENISA, 2023).

The MITRE ATT&CK's philosophy relies on the cyber security community to present the creation, enhancement, and abstraction of new TTPs with the aim of continually improving the taxonomy. The researcher proposes a potentially new sub-technique in relation to the deliberate damage of infrastructure which impact the availability of a network resource.

- Deliberate Damage—T1498.003

The creation of T1498.003—Deliberate Damage would encompass the sabotage of network infrastructure and can be attributed to multiple industries. Regarding the inclusion of subsea infrastructure this should reversed for nation-states who have the capability to conduct a cyber physical attack of this nature (ENISA, 2023).

- Maritime Taxonomy Creation

The data presented in the preceding section demonstrates a diverse range of adversarial strategies spanning thirteen of the fourteen ATT&CK tactics. This clearly demonstrates the maritime industry is not unique when faced with a cyber-attack and their adversary's employee similar techniques to other industries. This is reinforced by aligning the extraction of adversarial TTPs to the ATT&CK taxonomy. The analysis, quantification, and prioritisation provide cyber defenders with the appropriate and most effective detections and mitigations to be deployed within the maritime industry. A further seven reports relate to spoofing, jamming and physical sabotage of maritime infrastructure which were previously thought to be unique to the maritime industry but can be aligned with the ATT&CK taxonomy. The creation of a data frame, visualisation and critical analysis has allowed for an iterative approach initial perspective of the observed adversarial behaviours in relation to the maritime industry shown in Fig. 13.

- Taxonomy Gap

Privilege Escalation relates to an adversary gaining a higher-level permission on the system was not observed during the data extraction. The gap is evident in the data visualisation (11) and have not been reflected in the created taxonomy (13). Detecting privilege escalation requires extensive, monitoring and log analysis. Therefore, it

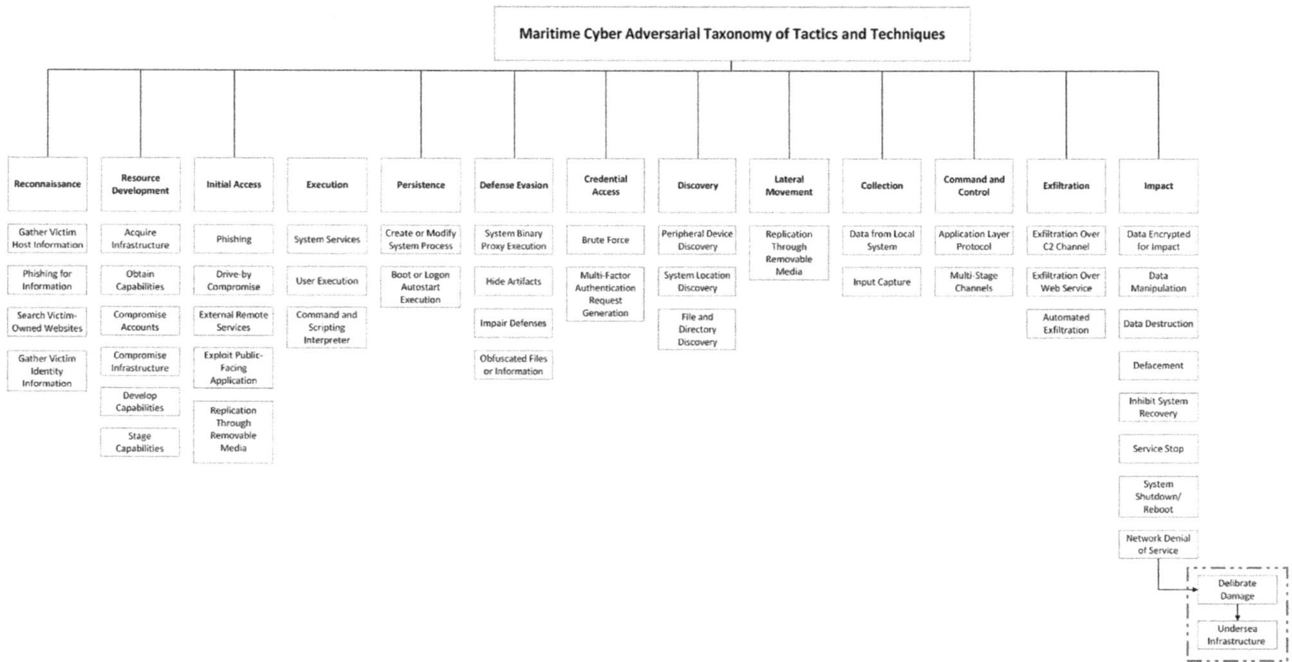

Fig. 13 Maritime taxonomy[4]

[4] The diagram was produced by the author with reference to (MITRE ATT&CK, n.d.) and the data contained in section **Error! Reference source not found.**

can be surmised that both onboard and ashore logging and monitoring capabilities were deficient. A contributing factor should be considered, had the detection capabilities been in place and the cyber-attack thwarted there is a requirement stated by BIMCO that the incident be reported. The guidance articulates "unauthorised or inappropriate use of administrative credentials" which aligns with privilege escalation should be reported to the company and considerations made for wider distribution with information sharing organisations (BIMCO, n.d.). The terminology used within this document is reflective of an insider threat and does not articulate the possibility of an external threat actor. The document aligns cyber security as a safety issue, culturally this has an impact as the sharing of safety issues may be considered detrimental especially when the terminology is articulated as a non-conformity. This present a perspective that the onboard crew are not following the correct procedures and may lead to a lack of reporting. These factors are supported by the lack of cyber security incidents available during the review of secondary data.

The Maritime Cyber Attack Database was launched in July 2023 (NHL Stenden, 2023). Their data is validated by supporting sources but refinement in relation to their inclusion and exclusion criteria would improve data quality. During the initial review it became apparent that the database contained an example of a cyber enabled crime, whereby a wireless camera was used for an illegal act onboard a vessel (Ceylan Yeginsu, 2023). This demonstrates the need to continually review physical security controls onboard but does not relate to a cyber-attack and should not be included.

7 Conclusions

The conclusion of this research has created a Maritime Cyber Adversarial Taxonomy of Tactics and Techniques which can be used to inform future policies, improve onboard architectures and visually present the complexities of cyber security to a non-technical audience. This research has demonstrated that a cyber-attack against the maritime industry will most likely commence via phishing which may impersonate a port official and contain a malicious link or attachment. This technique aligns with most common initial access vector of 2022 demonstrating that the industry is not uniquely targeted. The resultant impact will encrypt data rendering it unavailable and exposing the industry to financially motivated cyber criminals. Due to the converged and interconnected nature of the onboard architectures the consequences could be significant. The denial of a critical onboard safety system could result in a collision, grounding, or environmental damage. The onboard systems by design and regulation lack the basic cyber security controls aligned with confidentiality and integrity, which increases the cyber-attack surface. The research has extracted 53 unique adversarial techniques spanning 13 of the 14 MITRE ATT&CK tactics. The presentation of a data mathematically proves the hypothesis, that despite the remote nature of the maritime industry, it does not face a unique cyber adversary. Consideration that the maritime industry is more likely to be impacted by the geopolitical landscape is unique. Vessels regularly transit near contested territorial waters in support of their role but due to the technical vulnerabilities in relation to the safety equipment the tampering with the vessel positioning can impact the vessels safety and influence geopolitics. The production of a data visualisation aids in the engagement with non-technical audiences as they can easily identify areas of prioritisation based on empirical data which is relatable to the audience as previous incidents provide realism for positive change. The alignment of data manipulation in relation to GNSS and AIS spoofing, further enriches the ATT&CK taxonomy and the jamming of critical onboard safety signals presents a differing perspective on a denial-of-service attack. The Deliberate Damage of undersea infrastructure is proposed as an addition sub technique to the MITRE ATT&CK taxonomy as this aligns with the denial of a network at the cyber physical level. Aligning the maritime industry with the MITRE ATT&CK taxonomy provides the tangible benefit of understating the adversarial capabilities which ensure the implementation of cyber defences are appropriate in relation to the cyber threat and risk.

The future work aims to repeat the process by expanding the primary sources to encompass multiple languages. This would generate a further enriched dataset for evaluation of the geopolitical nature of the industry. This first phase would create an enriched adversarial model by reinforcing the identification of keystone techniques which would allow for a deeper technical output in relation to mitigations, detections, and defensive opportunities. The second phase would focus on the development and implementation of an enhanced onboard logging and monitoring capability utilising machine learning and artificial intelligence to aid the detection capabilities in the converged IT and OT environment.

The maritime industry is on a journey of recognition. The perception that operating in remote parts of the globe are preventive measures for a cyber-attack are not tangible due to the widespread interconnectivity of ship to ship and ship to shore communications. This research has suggested positive improvements of the onboard architectures and demonstrated the benefits adversarial modelling which would significantly benefit the maritime industry.

References

Applebaum, A., Nickels, K., Pennington, A., Schulz, T., Strom, B., & Wunder, J. (n.d.). Getting Started with ATT&CK.

Ashraf, I., Park, Y., Hur, S., Kim, S. W., Alroobaea, R., Zikria, Y. B., & Nosheen, S. (2023). A Survey on Cyber Security Threats in IoT-Enabled Maritime Industry. *IEEE Transactions on Intelligent Transportation Systems, 24*, 2677–2690. https://doi.org/10.1109/TITS.2022.3164678

ATT&CK® Navigator [WWW Document]. (n.d.). URL https://mitre-attack.github.io/attack-navigator/. Accessed November 18, 2023.

Australian Cyber Security Centre (ACSC). (n.d.). *ACSC Prevention and Protection Guide.*

Bank of England. (2016). *CBEST Intelligence-Led Testing Understanding Cyber Threat Intelligence Operations.*

Barrett, D., & Weiss, M. (2017). *CompTIA Security + SY0–501* (5th ed.). Que Corporation.

Bhattacharjee, S. (2021). *What is Electronic Chart Display and Information System (ECDIS)?* [WWW Document]. https://www.marineinsight.com/marine-navigation/what-is-electronic-chart-display-and-information-system-ecdis/. Accessed September 10, 2023.

BIMCO (n.d.). The Guidelines On Cyber Security Onboard Ships Produced and supported by the Guidelines on Cyber Security Onboard Ships.

Bjorn Bergman. (2021). *Systematic Data Analysis Reveals False Vessel Tracks—SkyTruth* [WWW Document]. https://skytruth.org/2021/07/systematic-data-analysis-reveals-false-vessel-tracks/. Accessed December 9, 23.

British Standards Institute. (2020). *BS EN 61162–1:2016: Maritime Navigation and Radiocommunication Equipment and Systems.* Digital Interfaces: Single Talker and Multiple Listeners.

British Standards Institute. (2015). *BS EN 61174:2015: Maritime Navigation and Radiocommunication Equipment and Systems.* Electronic Chart Display and Information System (ECDIS). Operational and Performance Requirements, Methods of Testing and Required Test Results.

British Standards Institute. (2003). *BS EN 60945:2002 Maritime Navigation and Radiocommunication Equipment and Systems.* General Requirements. Methods of Testing and Required Test Results.

Cabinet Office. (2023). *Professional Development Framework for all Source Intelligence Analysis.*

Caprolu, M., DI Pietro, R., Raponi, S., Sciancalepore, S., & Tedeschi, P. (2020). Vessels Cybersecurity: Issues, Challenges, and the Road Ahead. *IEEE Communications Magazine, 58*, 90–96. https://doi.org/10.1109/MCOM.001.1900632

Ceylan, Y., (2023). *Video Voyeur Hid Camera in Cruise Ship Bathroom, F.B.I. Says—The New York Times* [WWW Document]. https://www.nytimes.com/2023/05/10/travel/hidden-camera-cruise-ship-bathroom-fbi.html. Accessed December 9, 2023.

CISA. (2023a). *CISA Releases Decider Tool to Help with MITRE ATT&CK Mapping | CISA* [WWW Document]. https://www.cisa.gov/news-events/alerts/2023/03/01/cisa-releases-decider-tool-help-mitre-attck-mapping. Accessed October 26, 2023.

CISA. (2023b). GitHub—Cisagov/Decider: A Web Application that Assists Network Defenders, Analysts, and Researchers in the Process of Mapping Adversary Behaviors to the MITRE ATT&CK® Framework. [WWW Document]. https://github.com/cisagov/decider. Accessed November 19, 2023.

Clarke, T. (2023). Royal Navy's Experimental Ship Carries Out First Trial of Quantum Navigation System [WWW Document]. https://news.sky.com/story/royal-navys-experimental-ship-carries-out-first-trial-of-quantum-navigation-system-12889487. Accessed September 24, 2023.

Claudio, A., Stephen, C., Koen Van Impe, & Andreas, S. (2022). *ENISA Threat Landscape 2022.* https://doi.org/10.2824/764318

Collier, J., Ronis, S., Lane, I., & Simpson, R. (2023). *A Requirements-Driven Approach to Cyber Threat Intelligence.*

CompTIA (n.d.). *What Is Spoofing | Cybersecurity | CompTIA* [WWW Document]. https://www.comptia.org/content/articles/what-is-spoofing. Accessed December 9, 23.

Darktrace (n.d.). *Troubled Waters: Cyber-attacks on San Diego and Barcelona's Ports* [WWW Document]. https://darktrace.com/blog/troubled-waters-cyber-attacks-on-san-diego-and-barcelonas-ports. Accessed September 16, 2023.

Direnzo, J., Goward, D. A., & Roberts, F. S. (n.d.). *The Little-known Challenge of Maritime Cyber Security**.

DNV. (2023). *Maritime Cyber Priority 2023: Staying Secure in an Era of Connectivity* [WWW Document]. https://www.dnv.com/cybersecurity/cyber-insights/maritime-cyber-priority-2023.html?utm_source=google&utm_medium=cpc&utm_campaign=Awareness-DSA&gad=1&gclid=EAIaIQobChMIxIyXtauHggMVhYpQBh1NyAcxEAAYASAAEgIQmfD_BwE. Accessed October 21, 2023.

Dunning, H. (2023). Quantum Sensor for a Future Navigation System Tested Aboard Royal Navy Ship [WWW Document]. https://www.imperial.ac.uk/news/245114/quantum-sensor-future-navigation-system-tested/. Accessed September 24, 2023.

Edgar, T. W., & Manz, D. O. (2017). Addressing the Adversary. *Research Methods for Cyber Security, 345–366*. https://doi.org/10.1016/B978-0-12-805349-2.00014-5
Google (n.d.) *Ranking Results—How Google Search Works* [WWW Document]. https://www.google.com/intl/en_uk/search/howsearchworks/how-search-works/ranking-results/#. Accessed December 9, 2023.
Gregor, G. (2022). Police Suspect Jammer Devices on HGVs Caused Ferry's GPS system failure | trans.info [WWW Document]. https://trans.info/en/autonomous-trucking-company-raises-50m-funding-364720. Accessed December 9, 2023.
Harish, A. V. (2022). *Investigating the Security and Accessibility of Voyage Data Recorder Data using a USB attack Harish, Avanthika Vineetha Investigating the Security and Accessibility of Voyage Data Recorder Data using a USB Attack.*
Helmick, J. S. (2020). In-transit Threats and Risk Management. *Intermodal Maritime Security: Supply Chain Risk Mitigation,* 259–280. https://doi.org/10.1016/B978-0-12-819945-9.00022-8
Holub, A. (2023). *Threat Spotlight: STRRAT, ZLoader, and HoneyGain—Cisco Umbrella* [WWW Document]. https://umbrella.cisco.com/blog/cybersecurity-threat-spotlight-strrat-zloader-honeygain. Accessed October 26, 2023.
Hutchins, E. M., Cloppert, M. J., Amin, R. M. (n.d.). *Intelligence-Driven Computer Network Defense Informed by Analysis of Adversary Campaigns and Intrusion Kill Chains.*
IBM. (n.d.). *What is Data Visualization? | IBM* [WWW Document]. https://www.ibm.com/topics/data-visualization. Accessed June 25, 2023.
International Maritime Organization. (1996). *Resolution MSC.64(67)—Adoption of New and Amended Performance Standards.*
Kavallieratos, G., Diamantopoulou, V., & Katsikas, S. K. (2020). Shipping 4.0: Security Requirements for the Cyber-Enabled Ship. *IEEE Trans Industry Inform, 16*, 6617–6625. https://doi.org/10.1109/TII.2020.2976840
Kavallieratos, G., & Katsikas, S. (2020). Managing Cyber Security Risks of the Cyber-enabled Ship. *Journal of Marine Science and Engineering, 8*, 1–19. https://doi.org/10.3390/jmse8100768
Louart, M., Szkolnik, J. J., Boudraa, A. O., Le Lann, J. C., & Le Roy, F. (2023). Detection of AIS Messages Falsifications and Spoofing by Checking Messages Compliance with TDMA Protocol. *Digital Signal Process, 136*, 103983. https://doi.org/10.1016/J.DSP.2023.103983
Maillé, P., Maudet, G., Simon, M., & Tuffin, B. (2022). Are Search Engines Biased? Detecting and Reducing Bias using Meta Search Engines. *Electronic Commerce Research and Application,* 101132. https://doi.org/10.1016/J.ELERAP.2022.101132
Maritime and Coastguard Agency. (2013). *Marine Information Note Min 445 (M+F) Navigation Safety: Electronic Chart Display and Information System (ECDIS)-Use of ECDIS as Primary Means of Navigation (PMN).*
Maritime and Coastguard Agency. (2020). *Marine Guidance Note 610 (M+F) Navigation: SOLAS Chapter V-Guidance on the Merchant Shipping (Safety of Navigation) Regulations 2020.*
M-CERT (n.d.). *Maritime Computer Emergency Response Team (M-CERT)* [WWW Document]. https://www.m-cert.fr/admiral/. Accessed December 9, 2023.
Meland, P. H., Bernsmed, K., Wille, E., Rødseth, J., & Nesheim, D. A. (2021). A Retrospective Analysis of Maritime Cyber Security Incidents. *TransNav 15*, 519–530. https://doi.org/10.12716/1001.15.03.04
Meng, L., Yang, L., Yang, W., & Zhang, L., (2022). A Survey of GNSS Spoofing and Anti-Spoofing Technology. *Remote Sensing, 14*, 4826 14, 4826. https://doi.org/10.3390/RS14194826
MITRE (n.d.). *Groups | MITRE ATT&CK®* [WWW Document]. https://attack.mitre.org/groups/. Accessed June 18, 2023.
MITRE ATT&CK (2020a). *Replication Through Removable Media, Technique T0847 - ICS* [WWW Document]. URL https://attack.mitre.org/techniques/T0847/. Accessed September 23, 2023.
MITRE ATT&CK (2020b). *Communication Through Removable Media, Technique T1092 - Enterprise* [WWW Document]. URL https://attack.mitre.org/techniques/T1092/. Accessed September 23, 2023.

MITRE ATT&CK. (2017a). *External Remote Services, Technique T1133* [WWW Document]. https://attack.mitre.org/techniques/T1133/. Accessed September 23, 2023.
MITRE ATT&CK. (2017b). *Replication Through Removable Media, Technique T1091—Enterprise* [WWW Document]. https://attack.mitre.org/techniques/T1091/. Accessed September 23, 2023.
MITRE ATT&CK. (n.d.). *Network Sniffing, Technique T0842—ICS* [WWW Document]. https://attack.mitre.org/techniques/T0842/. Accessed September 23, 2023a.
MITRE ATT&CK. (n.d.). *Matrix—Enterprise | MITRE ATT&CK®* [WWW Document]. https://attack.mitre.org/matrices/enterprise/. Accessed October 6, 2023b.
MITRE ATT&CK. (n.d.). *ICS Matrix | MITRE ATT&CK®* [WWW Document]. https://attack.mitre.org/matrices/ics/. Accessed October 6, 2023c.
MITRE ATT&CK. (n.d.). *Techniques—Enterprise | MITRE ATT&CK®* [WWW Document]. https://attack.mitre.org/techniques/enterprise/. Accessed October 5, 2023d.
MITRE ATT&CK. (n.d.). *Techniques—ICS | MITRE ATT&CK®* [WWW Document]. https://attack.mitre.org/techniques/ics/ Accessed October 5, 2023e.
MITRE ATT&CK. (n.d.). *Phishing, Technique T1566—Enterprise* [WWW Document]. https://attack.mitre.org/techniques/T1566/ Accessed November 25, 2023f.
MITRE ATT&CK. (n.d.). *Replication Through Removable Media, Technique T1091—Enterprise* [WWW Document]. https://attack.mitre.org/techniques/T1091/ Accessed December 2, 2023g.
MITRE ATT&CK. (n.d.). *Data Encrypted for Impact, Technique T1486—Enterprise | MITRE ATT&CK®* [WWW Document]. https://attack.mitre.org/techniques/T1486/. Accessed December 10, 2023h.
MITRE ATT&CK. (n.d). *Network Denial of Service, Technique T1498—Enterprise | MITRE ATT&CK®* [WWW Document]. https://attack.mitre.org/techniques/T1498/. Accessed December 9, 23i.
MITRE ATT&CK. (n.d.). *Data Encrypted for Impact, Technique T1486—Enterprise | MITRE ATT&CK®* [WWW Document]. https://attack.mitre.org/techniques/T1486/. Accessed December 11, 2023j.
NCSC. (2023). *20th Anniversary of First Response to State-Sponsored Cyber Attack* [WWW Document]. https://www.ncsc.gov.uk/news/20th-anniversary-of-first-response-to-state-sponsored-cyber-attack. Accessed September 2, 2023.
NCSC (2018). *Introduction to Logging for Security Purposes* [WWW Document]. https://www.ncsc.gov.uk/guidance/introduction-logging-security-purposes. Accessed November 26, 2023.
NHL Stenden (2023). *Researchers at NHL Stenden Launch Database that Exposes Cyber Hacking in the Worldwide Maritime Industry | NHL Stenden University of Applied Sciences* [WWW Document]. https://www.nhlstenden.com/en/news-and-articles/researchers-at-nhl-stenden-launch-database-that-exposes-cyber-hacking-in-the. Accessed December 9, 2023.
NHL Stenden University. (n.d.). *Maritime Cyber Attack Database (MCAD)* [WWW Document]. https://www.nhlstenden.com/en/maritime-cyber-attack-database. Accessed November 12, 2023.
NPSA. (2023). *Critical National Infrastructure | NPSA* [WWW Document]. https://www.npsa.gov.uk/critical-national-infrastructure-0. Accessed September 16, 2023.
OECD. (2016). The Ocean Economy in 2030. *Directorate for Science, Technology and Innovation Policy Note.* https://doi.org/10.1787/9789264251724-en
OFCOM. (2016). *VHF Radio Spectrum for the Internet of Things Statement Spectrum for the Internet of Things.*
OPSWAT. (2018). *What Does "Sheep Dip" Mean in Cyber Security?—OPSWAT* [WWW Document]. https://www.opswat.com/blog/what-does-sheep-dip-mean-cyber-security. Accessed December 2, 2023.
Pols Paul. (2023). *The-Unified-Kill-Chain.*
Regenscheid, A. (2018). *Platform Firmware Resiliency Guidelines.* https://doi.org/10.6028/NIST.SP.800-193
Rick Vanover. (2021). *What is the 3–2–1 Backup Rule?* [WWW Document]. https://www.veeam.com/blog/321-backup-rule.html. Accessed December 10, 2023.

Saunders, M. N. K. (2023). *Research Methods for Business Students* (9th ed.). Pearson.
Schinas, O., & Metzger, D. (2023). Cyber-seaworthiness: A Critical Review of the Literature. *Marine Policy, 151*. https://doi.org/10.1016/j.marpol.2023.105592
Shabashnyi, H. (2015). Gaining the Advantage Applying Cyber Kill Chain Methodology to Network Defense. *Shipping 4.0*, 2016, in: Maritime Week—SINTEF.
Silverajan, B., Ocak, M., & Nagel, B. (2018). Cybersecurity Attacks and Defences for Unmanned Smart Ships. In *2018 IEEE International Conference on Internet of Things (IThings) and IEEE Green Computing and Communications (GreenCom) and IEEE Cyber, Physical and Social Computing (CPSCom) and IEEE Smart Data (SmartData)* (pp. 15–20). IEEE. https://doi.org/10.1109/Cybermatics_2018.2018.00037
Snowball sampling [WWW Document]. (n.d.). https://research-methodology.net/sampling-in-primary-data-collection/snowball-sampling/. Accessed October 21, 2023.
Stouffer, K., Pillitteri, V., Lightman, S., Abrams, M., & Hahn, A. (2015). *Guide to Industrial Control Systems (ICS) Security*. https://doi.org/10.6028/NIST.SP.800-82R2
Stringhini, G., Joinson, A., & Taylor, P. (n.d.). *Adversarial Behaviours Knowledge Area.*
Strom, B. E., Applebaum, A., Miller, D. P., Nickels, K. C., Pennington, A. G., & Thomas, C. B. (2020). *MITRE ATT&CK: Design and Philosophy.*
ENISA. (2023). Subsea Cables-What Is At Stake? Subsea Cables-What Is At Stake? *About ENISA*. https://doi.org/10.2824/212261
Suzuki, M., & Yamamoto, Y. (2021). Characterizing the Influence of Confirmation Bias on Web Search Behavior. *Frontiers in Psychology, 12*. https://doi.org/10.3389/FPSYG.2021.771948/FULL
Svilicic, B., Rudan, I., Jugović, A., & Zec, D. (2019). A Study on Cyber Security Threats in a Shipboard Integrated Navigational System. *Journal of Marine Science and Engineering, 7*. https://doi.org/10.3390/jmse7100364
Tam, K., & Jones, K. (n.d.). *Cyber-Risk Assessment for Autonomous Ships.*
Toulas, B. (2022). *Phishing Impersonates Shipping Giant Maersk to Push STRRAT Malware* [WWW Document]. https://www.bleepingcomputer.com/news/security/phishing-impersonates-shipping-giant-maersk-to-push-strrat-malware/. Accessed October 26, 2023.
Trend Micro. (n.d.). *Indicators of Compromise—Definition* [WWW Document]. https://www.trendmicro.com/vinfo/us/security/definition/indicators-of-compromise. Accessed October 4, 2023.
United States Coast Guard Cyber Command. (2022). *2022 Cyber Trends and Insights in the Marine Environment.*
USCG. (2019). *Cyber Incident Exposes Potential Vulnerabilities Onboard Commercial Vessels.*
USCG. (n.d.). *Coast Guard Maritime Commons* [WWW Document]. https://www.news.uscg.mil/maritime-commons/Category/23882/cyber-awareness-risk-management/. Accessed November 12, 2023.
Weaver, G. A., Feddersen, B., Marla, L., Wei, D., Rose, A., & Van Moer, M. (2022). Estimating Economic Losses from Cyber-Attacks on Shipping Ports: An Optimization-Based Approach. *Transportation Research Part c: Emerging Technologies, 137*. https://doi.org/10.1016/j.trc.2021.103423
White, T. (2019). *Cyber Threat Intelligence in Government: A Guide for Decision Makers & Analysts.*
Wolsing, K., Saillard, A., Bauer, J., Wagner, E., Van Sloun, C., Fink, I. B., Schmidt, M., Wehrle, K., & Henze, M. (n.d.). *Network Attacks Against Marine Radar Systems: A Taxonomy, Simulation Environment, and Dataset*. https://doi.org/10.5281/zenodo.6805559

Enhancing Security of Unmanned Aerial Vehicle Through Vulnerability Assessment and Penetration Testing: A Case Study on Parrot AR Drone 2.0

Akinola Aderinto, Sina Pournouri, and Sohrab Moshiri

Abstract As Unmanned Aerial Vehicles (UAVs), or drones, increasingly permeate future societal frameworks, safeguarding them from malign entities becomes crucial for their integration into essential daily operations. Forecasts by (Stanković et al., Drones 5:49, 2021) suggest the drone market will swell to USD 60 billion by 2025, signifying their expansive adoption and the technological advancements they bring to various industries (Hamdi et al., Drone Forensics: A Case Study on DJI Phantom 4. In 2019 IEEE/ACS 16th International Conference on Computer Systems and Applications [AICCSA], 2019). However, the widespread deployment of UAVs also escalates security dilemmas, such as unauthorized access and cyber-physical attacks (Bouafif et al., Drone Forensics: Challenges and New Insights. In 2018 9th IFIP International Conference on New Technologies, Mobility and Security [NTMS], 2018). The Parrot AR Drone 2.0, celebrated for its affordability and simplicity, is particularly prone to exploits that threaten data's confidentiality, integrity, and availability (Astaburuaga et al., Vulnerability Analysis of AR.Drone 2.0, an Embedded Linux System. In 2019 IEEE 9th Annual Computing and Communication Workshop and Conference [CCWC], 2019). This research targets these vulnerabilities through detailed vulnerability assessment and penetration testing, focusing on identifying and mitigating security flaws in the Parrot AR Drone 2.0's communication protocols and its unsecured FTP and Telnet connections. The findings aim to contribute to the UAV security domain, ensuring the safe integration of drones into critical infrastructures.

Keywords Parrot AR Drones 2.0 · Drone · UAV · Linux kernel · Busybox · Vulnerabilities · Risk · Vulnerability assessment · Penetration testing · Hacking · Countermeasures

---

A. Aderinto · S. Pournouri (✉) · S. Moshiri
Sheffield Hallam University, Sheffield, UK
e-mail: S.Pournouri@shu.ac.uk

H. Jahankhani et al. (eds.), *Space Governance*, Space Law and Policy,
https://doi.org/10.1007/978-3-031-62228-1_3

1 Background

Unmanned Aerial Vehicles (UAVs), also known as drones, have witnessed a surge in popularity due to their multifaceted utility across a broad spectrum of applications, from agriculture and construction to surveying, mapping, transportation, and defence, underscoring their transformative potential across industries (Yaacoub et al., 2020b). This rapid proliferation, however, has introduced a complex array of security challenges, including unauthorized access, data theft, and cyber-physical attacks, necessitating a reevaluation of their security frameworks (Kumar & Mohanty, 2021).

1.1 Common Security Threats to Drones

As with any interconnected technology, drones are susceptible to a spectrum of threats and vulnerabilities, magnified by their escalating popularity. This appeal extends beyond legitimate users, drawing the focus of malicious entities. The ensuing analysis delineates prevalent security challenges associated with drone operations:

- **Unauthorized Access and Control**
  There exists a tangible risk of adversaries exploiting drone communication protocols or leveraging weak authentication systems to usurp control, potentially resulting in the unauthorized use or commandeering of these devices (Yaacoub et al., 2020a, b).
- **Data Interception**
  Given their role in capturing and relaying multimedia and other forms of data, drones' transmissions, if insufficiently protected, are prone to interception. Such breaches can compromise privacy and result in the disclosure of confidential information (Siddiqi et al., 2022).
- **GPS Spoofing and Jamming**
  Drones' dependency on GPS for navigation makes them vulnerable to signal spoofing and jamming, tactics that can mislead or destabilise their operational trajectory (Kumar & Mohanty, 2021).
- **Malware and Firmware Manipulation**
  Analogous to other IoT devices, drones can fall prey to malware, with inadequately secured firmware being a vector for malicious code injection. This can facilitate unauthorized access, data exfiltration, or physical compromise of the drone (Shafique et al., 2021).
- **Physical Attacks**
  The tangible nature of drones also exposes them to direct physical threats, which can be particularly disruptive to those engaged in surveillance, delivery, or essential services. Such assaults may target the drone's structural integrity, sensory apparatus, or operational pathways (Amrami, 2023).

- **Technical Problems**
  Operational reliance on battery systems introduces risks, with flaws in battery management potentially being exploited to induce control loss or safety hazards (Siddiqi et al., 2022).
- **Communication Security Flaws**
  Inadequate encryption within drone communications amplifies the risk of data interception and manipulation, underscoring the need for robust cryptographic practices.
- **Firmware Maintenance Gaps**
  The absence of regular firmware updates can leave drones exposed to known vulnerabilities and security defects, emphasising the importance of timely software maintenance.
- **Privacy Intrusions**
  The use of drones for imaging or surveillance, without appropriate consent or oversight, engenders significant privacy and ethical dilemmas, necessitating stringent operational guidelines.

1.2 Related Works

Drones, defined as remotely piloted or autonomously operated aircraft, vary widely in design from compact quadcopters to expansive fixed-wing models, reflecting their diverse operational capabilities (Hamdi et al., 2019; Shafique et al., 2021). The AR Drone 2.0, a consumer-grade model favored for its cost-effectiveness, user-friendliness, and innovative features, notably features dual cameras facilitating live video streaming over WLAN. While these attributes enhance operational control, they also present exploitable avenues for security and privacy violations.

The integration of drones into daily operational contexts amplifies the urgency of addressing their security vulnerabilities. The spectrum of potential security risks encompasses unauthorised system access, data breaches, and physical tampering, highlighting the critical need for fortified security measures (Iqbal et al., 2019). Vulnerabilities within drone firmware, network communications, mobile interfaces, and flight control systems present exploitable targets for malicious entities aiming to disrupt operations or compromise data integrity.

To mitigate these risks, conducting thorough vulnerability assessments and penetration testing on models such as the Parrot AR Drone 2.0 is imperative. Such analysis not only elucidate specific security weaknesses but also foster the development of comprehensive security protocols, offering pivotal insights for drone manufacturers, operators, and regulatory bodies. Through detailed vulnerability exploration and impact assessment, this research aspires to advance the establishment of rigorous security standards, guidelines, and best practices, thereby enhancing the safety, reliability, and trustworthiness of drone operations.

Over recent years, the academic and professional security communities have dedicated considerable effort to unravelling the vulnerabilities inherent in Unmanned Aerial Vehicles (UAVs). Such endeavours are pivotal in delineating the security landscape of drone technology, aiming to unearth potential risks, weaknesses, and avenues for exploitation. This section synthesizes seminal works in the domain of drone vulnerabilities, distilling key findings and insights from a corpus of existing literature.

Bouafif et al. (2018) present a compelling forensic analysis of the Parrot AR Drone 2.0, uncovering avenues for unauthorized access to the drone's file system via serial or FTP connections. Their work notably elucidates the retrieval process of the controller's ID, a critical piece of information that can substantiate ownership claims for this specific drone model.

In contrast, the investigation by Hamza and Qassim (2021) into the DJI Mini 2's security posture reveals a robust resistance to eavesdropping attempts on its wireless communication channels (FHSS and OFDM), attributed to the encrypted nature of these links. This finding highlights the variable security profiles across different drone models and the importance of encryption in safeguarding drone communications.

Yaacoub et al. (2020a, b) embark on an exploration of drone vulnerabilities, identifying a spectrum of exploitable weaknesses within both military and civilian drones' communication frameworks. Their methodology, rooted in a simulated attack hacking cycle, underscores the complexity of drone security, exacerbated by the diverse communication and network protocols utilized by drones and associated smart devices.

Further expanding the discourse, Choudhary et al. (2018) and Kumar and Mohanty (2021) delve into the vulnerabilities of the Internet of Drones (IoD), echoing findings similar to those of Yaacoub et al. (2020a, b) concerning software and data transmission vulnerabilities. However, their reliance on literature review rather than empirical penetration testing points to a gap in hands-on vulnerability assessment in drone research.

The collective body of research on drone vulnerabilities articulates a clear imperative for enhanced security protocols. It stresses the necessity for secure communication channels, rigorous firmware and software development standards, and stringent authentication mechanisms. Leveraging the insights garnered from these studies, the security of UAVs can be significantly bolstered, mitigating potential threats and promoting the safe, secure, and responsible integration of drones across various sectors.

Recent scholarly investigations have delved into the security of remote-control protocols within Unmanned Aerial Vehicles (UAVs), uncovering significant vulnerabilities in drone-controller communications. These vulnerabilities encompass inadequate encryption, absence of authentication measures, and exposure to replay attacks. Astaburuaga et al. (2019) conducted a comprehensive analysis of firmware and software across several drone models, revealing critical security flaws such as insecure coding practices, buffer overflows, and the presence of backdoor accounts.

Further expanding the discourse, Yaacoub et al. (2020a, b) highlighted the potential for GPS signal manipulation as an attack vector. By altering the GPS signals received by UAVs, adversaries could potentially redirect flight paths or precipitate crashes, posing severe operational risks. Kharchenko and Torianyk (2018) and Marty (2014) shed light on the vulnerabilities inherent in wireless communication protocols, including Wi-Fi and radio frequencies, which are pivotal for remote control and data transmission.

Clark et al. (2017) underscored the threat posed by physical tampering, demonstrating that unauthorized modifications to drone hardware could facilitate access to sensitive data or enable the insertion of malicious components. Despite the depth of these studies in identifying various vulnerabilities, there remains a critical oversight concerning the evaluation of risk impacts that such vulnerabilities may have on individuals when exploited by malicious actors.

This research endeavours to bridge this gap by assessing the risk impact of vulnerabilities identified in the AR Drone 2.0 through comprehensive vulnerability assessments. Moreover, this study distinguishes itself by not only identifying but also mitigating several of these security vulnerabilities within the device. Consequently, this research contributes uniquely to the body of knowledge on UAV security, providing insights into both the identification and remediation of critical vulnerabilities in drone systems.

2 Methodology

In the realm of vulnerability detection and penetration testing, numerous methodologies can be employed. The results of every penetration testing might be different from each other depending on the methodology and standards chosen for the practice.

When conducting vulnerabilities assessment and penetration testing, it is important to know the areas to check in the device to obtain sensitive and confidential information. Most embedded systems consist of a single IT system. However, they are usually composed of a variety of components (CPU, sensors, memory card and the camera) that communicate with each other, like a network of different IT systems of a company. Those components also generate data, such as sensor data from the ultrasound altimeter, which the embedded system processes.

2.1 Research and Experiment Design

The significant preceding literature research that was undertaken served as the foundation behind the Parrot AR Drone 2.0 equipment selection. According to the research, this device is one of the most widely used, easily accessible to everyone without any prerequisites, and moderately priced. Another reason is the lack of a systematic approach for securing Parrot AR Drone 2.0.

After identifying the vulnerabilities present in the AR Drone 2.0, attacks were launched to compromise the CIA of the device. MITM attack was launched to hijack

Table 1 Hardware components for the experiment

Hardware	Description and utilisation
UAV	Parrot AR Drone 2.0
GCS	The GCS used in this research is an Apple product, which sends commands to the AR drone 2.0 through the AR.FreeFlight 2.4.15 application and waits for replies from UAV through telemetry
Host	The host is a HP Pavilion with 64-bits Windows 10 Operating System, 6 GB RAM, and of CPU i5-3210 M CPU @ 2.50 GHz
Attacker	A VirtualBox running a kali Linux OS was installed on a HP pavilion laptop with 6 GB RAM and 225 SSD
Network Monitor	Cisco network adapter was attached to the above setup (HP pavilion laptop with VB running a kali Linux)

the live stream video transmitting from the UAV to the controller. A successful breach of availability was attained because it disconnected the communication link between the UAV and the controller, knocking out the legitimate user and granting access to an unauthorised user. A breach of integrity was also successful with the process because the device was fed with a false information making it look legitimate.

A well-configured and controlled test environment is essential for conducting a successful and effective vulnerability assessment and penetration testing on the AR Drone 2.0. The test environment replicates the real-world conditions as closely as possible while ensuring the safety and integrity of both the drone system and the testers. A well-designed test environment ensures that the testing is conducted safely, without any harm to the drone or its surroundings.

Table 1 outline the complete list of the equipment used to develop the case study scenarios and carry out the experiment for the study.

2.2 *Vulnerability Assessment*

Vulnerability assessment involves identifying and documenting potential weaknesses and vulnerabilities in a system. It focuses on identifying security flaws, misconfigurations, or weaknesses in software, hardware, network configurations, or operational procedures. The AR Drone 2.0, as a specific drone model, has its own set of vulnerabilities that need to be addressed.

The results of the vulnerability scans were analysed, and vulnerabilities were prioritised based on their severity and potential impact. The potential risks associated with the identified vulnerabilities were assessed, considering their likelihood of exploitation and potential consequences.

Table 2 AR Drone 2.0 open ports and purpose

Open ports	Purpose
21 FTP	Transfers the media (video and image files) recorded by the drone
23 Telnet	Provides remote access and root shell to the drone
5551 (FTP)	Provides FTP access to the update folder for the purpose of BusyBox updates
5553 (Sgi-eventmond)	This provides the 720p video frames while the drone application is recording
5555 Freeciv	Provides the video streams recorded by the drone to the clients
5557 Farenet	Provides assistance to the controller and contains important data such as the configuration files
5559 (TCP)	Provides assistance to the controller and contains important data such as the configuration files

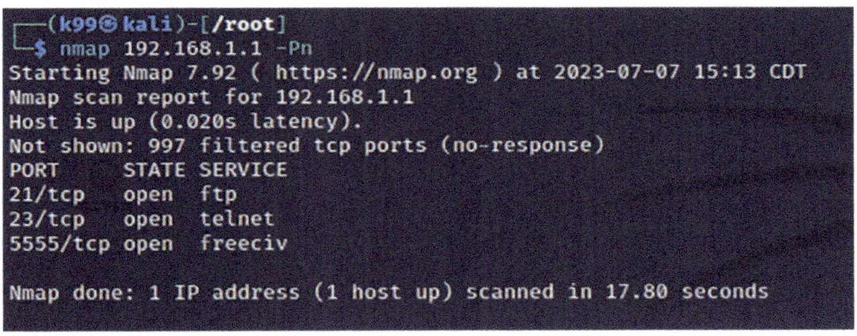

Fig. 1 Nmap result

2.2.1 Network Scan for Vulnerabilities

Nmap scan and Sparta show that some useful ports were open. These ports serve various functions, primarily utilised by the AR.FreeFlight application for drone control. While some of the additional ports are not involved in drone control, they provide assistance to the controller during drone updates. The communication protocol between the drone and the controller was further investigated using Wireshark. Some more useful information was captured as shown in Fig. 5. Nessus vulnerability scan came out with some critical vulnerabilities which will be further discussed in the analysis section (Table 2; Figs. 1, 2).

2.2.2 Result Analysis

To analyse the result of the network scan, we looked at the importance of those open ports and what might interest bad actors. Attackers can target a network or device using a variety of strategies and reasons. Attackers may target a UAV to eavesdrop

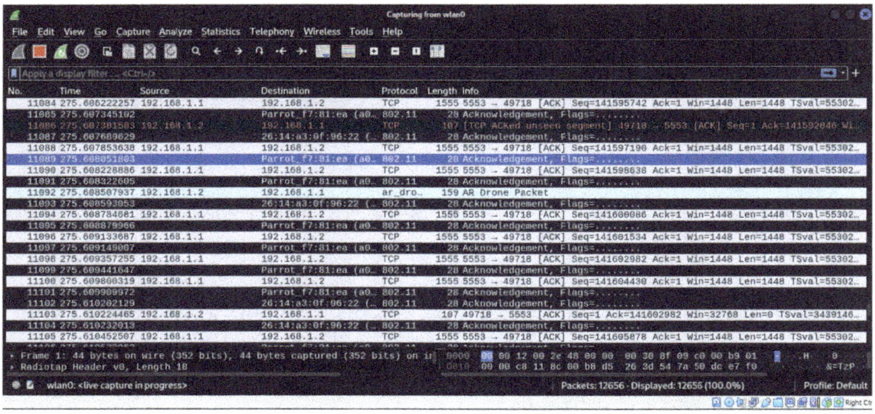

Fig. 2 Wireshark result

on it to learn what is being watched (this is known as eavesdropping), to hijack it and use it for analysis or another purpose (this is known as hijacking), or to stop ongoing aerial surveillance (this is known as denial-of-service).

From the results of the network scan, FTP, Telnet and Freeciv were found open. FTP is used for transferring images and videos recorded by the drone, and some other important files, which might interest bad actor to steal sensitive information. The likelihood of this service being exploited by bad actors is very high because of some information it might contain. This action violates the confidentiality and integrity of the device. Telnet on the other hand is a network protocol that allows users to remotely access and manage devices, computers, and servers over a network, such as the internet or a local area network (LAN).

Knowing what these ports are used for might attract malicious user to take control over the device. The likelihood of the port being exploited is very high which might lead to denial-of-service attack, in turn might leads to non-availability of services in each system.

2.2.3 Vulnerabilities Prioritisation

To prioritise the vulnerabilities, the likelihood of the vulnerability being exploited and the risk impact to the device or associated network must be known. Based on the Nessus vulnerability scanning results, it was shown that two of the vulnerabilities are critical if they were exploited. Telnet vulnerability is critical because it might lead to hijacking or denial-of-service of the device. FTP on the other hand is critical as this will lead to violation of system confidentiality and integrity (Fig. 3).

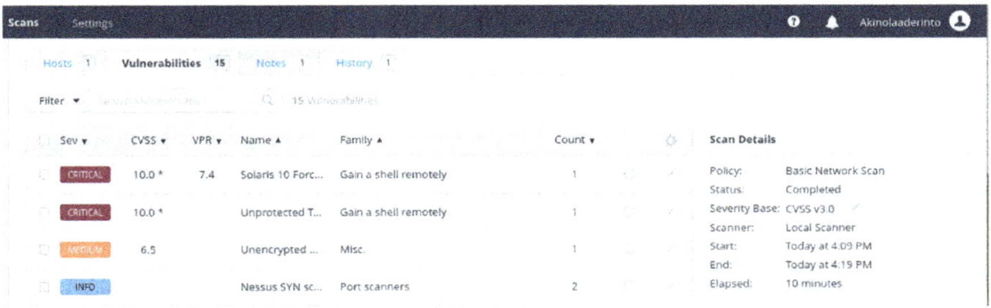

Fig. 3 Nessus scan

2.3 Penetration Testing

Penetration testing and vulnerability assessments are crucial for assessing the security of systems, particularly for Unmanned Aerial Vehicles (UAVs), to ensure their safe operation. This section outlines the streamlined penetration testing methodology applied in our research on UAVs, focusing on identifying and exploiting vulnerabilities in the Parrot AR Drone 2.0.

Methodology Overview:

1. **Reconnaissance and Scanning**: Initial steps involved using tools like Nmap and Sparta for identifying services and open ports, followed by Wireshark for analysing communication, and Nessus for detecting critical vulnerabilities.
2. **Enumeration and gaining Access**: Investigation revealed open FTP and Telnet services, allowing root access without authentication, highlighting severe security lapses.
3. **Privilege Escalation**: Examination of system vulnerabilities, software flaws, and weak security practices was conducted. Access to the device's "bin" directory revealed numerous commands available for root users, including potentially harmful "Destructive commands."
4. **Exploitation**: Leveraged root access to execute unauthorized actions, including uploading malicious files and executing a de-authentication attack via the "reboot" command, successfully disrupting the device-controller connection. Additionally, eavesdropping and Man-in-the-Middle (MITM) attacks were executed to further demonstrate the UAV's vulnerabilities.

This condensed approach emphasizes the critical findings and steps taken during the penetration testing process, highlighting significant security concerns within UAV systems that require immediate attention to prevent potential misuse or harm.

2.4 Data Collection

Data must be acquired by flying the Parrot AR Drone 2.0 in an appropriate location before analysis. In our experiment, we conducted data acquisition on various components of the Parrot AR Drone 2.0, including the drone itself and the controller. This drone relies on multiple systems to deliver its full range of functionalities. Our main objective was to acquire confidential information from the AR Drone 2.0 such as the file systems and memory, images and the video, the flight path, to run some destructive commands and to hijack the drone from the legitimate user. The emphasis was on compromising the CIA of the device while gathering this data.

The following detailed the process of our data acquisition:

- **Data from Wireshark**
  Wireshark was used to collect data from the communication protocol between the drone and the controller. The purpose of employing this was to monitor and evaluate the network data exchanged between the AR Drone 2.0 and its controller. This helps in the identification of potential communication protocol vulnerabilities and security concerns.
- **Data from Nmap scan and Sparta**
  According to the Nmap scan and Sparta, it has been observed that there are several open ports. The findings indicate that FTP, Telnet, and Freeciv port were operational and accessible. The ports serve various purposes and are primarily utilised by the AR.FreeFlight application for drone control and information transfer. These data were useful because it helps uncover accessible ports, services, and potential entryways within the drone's network.
- **Data from Nessus**
  The data obtained from the Nessus vulnerability scan helps to prioritise the vulnerabilities based on their criticalities. They produced reports regarding possible vulnerabilities that necessitate additional scrutiny. The full result was later discussed in Sect. 5 of this document.
- **Data from FTP**
  After a successful connection, two directories with root access were listed on the FTP. While exploring FTP directories, we had the capability to access and view any images stored on the drone, if there were indeed files, photos, or videos available.
- **Data from Telnet**
  Telnet is where we acquired the largest data. From the Telnet, we were able to acquire a lot of useful information, such as the Operating system, the firmware, the MAC address, the Linux kernel version, what type of privilege we had, the processes running on the device and so on.

  To gain knowledge about the available commands that users can execute on the device, we accessed the "bin" directory to obtain a comprehensive list of commands accessible after obtaining root access to the AR Drone 2.0 system. It is important to note that certain commands, referred to as "Destructive commands," possess the capability to inflict harm or injury to individuals when utilized by malicious actors.

2.5 Data Analysis

This section details the analysis of the data collected during the experiment. The result of the analysis shows that the research has met its aims by finding the vulnerabilities associated with the AR Drone 2.0 and providing an effective recommendation for mitigation.

Data collected from the vulnerability assessment and penetration testing were analysed to identify patterns and common vulnerabilities. The analysis helped in understanding the severity and impact of vulnerabilities and inform the development of effective security strategies.

The findings from all the discoveries made during the penetration testing evaluation, encompassing the identified vulnerabilities, instances of successful exploitation, and any sensitive information that was accessed throughout the process were compiled.

Severity Assessment of each identified vulnerability based on its potential impact on the drone system was carried out. Metrics like the CVSS from Nessus scan results was also considered to objectively rank vulnerabilities according to their risk levels.

We examine the possible repercussions that may arise from the effective exploitation of each vulnerability. We also take into account the ramifications of unauthorised entry, exposure of data, control of the drone, or any other detrimental impact on the operations of the drone and its users.

During the penetration testing assessment, it is imperative to analyse the attack scenarios utilised to comprehend the techniques employed by the researchers in exploiting vulnerabilities. This examination aids in the identification of the most probable attack vectors and potential areas of compromise.

To promptly address critical security issues, we prioritised the identified vulnerabilities according to their severity, impact, and exploitability. Our primary focus is on resolving high-risk vulnerabilities.

Here's how to approach test results and analysis:

- **Telnet Analysis**

  To gain full control over the drone, we established a telnet connection with the drone by connecting to its network. This connection granted us with root access, as indicated by the absence of a password for the root account in the /etc/passwd file. This unrestricted access enabled us to modify various system components, including binary, script, and configuration files.

  The feature also granted us the capability to execute various commands on the drone, such as halt, reboot, and shutdown. These commands effectively terminate the drone's normal operation, potentially leading to a sudden drop from the sky and subsequent damage. The likelihood of the port being exploited is very high which might lead to denial-of-service attack.

- **FTP Analysis**

  The drone is equipped with an open FTP server that offer unrestricted access. This server allows anonymous users to read from and write to them. Specifically, these servers grant access to the /update/ folder, which is used for firmware uploads, and the /data/video/ folder, which stores camera footage. Originally, this server was designed for uploading firmware updates and downloading camera footage.

  However, they can also be utilised to facilitate file transfers to and from the drone. Moreover, by editing the /etc/inetd.conf file via telnet, the entire system can be hosted, providing complete access to all files through the FTP server. The likelihood of this service being exploited by bad actors is very high because of some information it might contain.

- **Communications Analysis**

  The Parrot AR Drone 2.0 makes use of Open Wi-Fi as a means of communication between the controller and the drone. To analyse the communication between the AR Drone 2.0 and the controller, we first identified the computing devices on the network. AR Drone access point was first identified and connected to. No authentication was required to connect the drone and the controller. The IP addresses were also identified.

  The Wi-Fi adapter was put into monitor mode to analyse or eavesdrop the communication on the network. Wireshark was the tool used to listen to the communication signals. The port 5555 Freeciv for transmitting video signals was identified during eavesdropping.

- **System Analysis**

  The result of the Nmap scan showed that there are some open ports left on the system which is not ideal. The result of the Nmap scan showed that the ftp and telnet port were left open, also the port 5555 for live streaming of video was left open. Ftp allows for file sharing while Telnet helps for remote access of the device while the Freeciv is for video streaming leaving these ports vulnerable to potential attacks. To verify the accessibility of port 23, the Telnet command was employed, resulting in a root shell being obtained without requiring a password. Table 3 provides the details of open port obtained through Nmap scan.

By performing a thorough analysis of the test results, the security weaknesses in the AR Drone 2.0 can be better understood, and appropriate measures can be taken to enhance the drone's overall security posture. The analysis serves as a foundation for effective risk management and the development of proactive security measures to protect the drone system from potential threats and attacks. Table 3 shows the files and scripts in AR Drone 2.0

Table 3 Files and scripts in AR Drone 2.0

Directory	Information contains
/data/syslog.bin	This is a system log file that contains details of several events from AR Drone 2.0 operation system such as the file creation logs, configuration data and mount and version information
/data/config.ini	It contains the configuration files for the AR Drone 2.0 which includes the serial number, software version, drone name and Access point SSID
/data/emergency.bin	
/data/custom.configs/sessions/	This file contains AR Drone 2.0 GPS data
bin/program.elf	The main software that governs and controls the system
/bin/check_update.sh	This verifies if there are any updates accessible for the device
/bin/wifi_setup.sh	It helps to establish the WiFi device and generate an Access Point (AP)
/bin/reset_config.sh	It helps in configuration file reset and maintains total flight time

2.6 Evaluation

This section provides security evaluation of the Parrot AR Drone 2.0 based on the experiment carried out on the device. The result of the experiment shows that Parrot AR Drone 2.0 is vulnerable to several attacks which violate the CIA of information being processed and stored in the device.

The AR Drone 2.0 demonstrated vulnerabilities in its communication protocols, including weak encryption and susceptibility to man-in-the-middle attacks, weak authentication mechanisms, allowing unauthorised access to the drone's controls and data. We identified vulnerabilities within the drone's firmware, including potential code execution vulnerabilities and a lack of secure update mechanisms.

The authentication weaknesses identified in the AR Drone 2.0 can have a severe impact. Unauthorised access can lead to complete control of the drone, including navigation and camera functions, potentially compromising user privacy and data security.

Vulnerabilities in UAV communication systems are often rated as high severity. Potential impacts include unauthorised access, data interception, and manipulation of commands or telemetry data. Firmware and software vulnerabilities are frequently rated as high severity due to their potential to compromise the entire UAV system. Privacy-related vulnerabilities are often rated based on their potential to violate privacy regulations. Unauthorised data capture, especially if it includes personally identifiable information, can have significant legal and reputational consequences (Table 4).

Table 4 Security aspect requirements for AR Drone 2.0 components

Components	CIA Triad	Protection requirement	Reason
WLAN	Confidentiality Integrity Availability	Very High	Unencrypted Communication link
FTP	Confidentiality Integrity Availability	Very High	No authentication
Telnet	Confidentiality Integrity Availability	Very High	No authentication
Gyroscope	Availability	High	Contains information of drone stability to maintain a reference direction
Accelerometer	Availability	High	Contains information of drone stability and acceleration to maintain a reference direction
GPS	Confidentiality Availability	Very High	Contains drone's location information and some pre-processed data
RAM	Confidentiality Integrity Availability	Very High	Contains several sensitive and confidential information such as images and videos, some other data that are processed by the drone

2.7 Ethical Considerations

Throughout the research, ethical considerations were given utmost importance. The research adhered to ethical guidelines, ensuring participant confidentiality, informed consent, and protection of sensitive information. Any potential risks associated with the penetration testing or data collection were carefully assessed and mitigated. Although the device (Parrot AR Drone 2.0) used in this research was owned by the researchers. The assessment activities were non-destructive and do not cause harm to the Parrot AR Drone 2.0 or its associated systems.

The vulnerability assessments and penetration testing activities were in compliance with all applicable laws and regulations and the research activities did not violate any intellectual property rights, privacy laws, or other legal frameworks. All data were securely handled, strict confidentiality and privacy of any sensitive information or data obtained during the research was maintained.

Table 5 AR Drone 2.0 threat modeling

Threat type	Analysis
Spoofing	AR.FreeFlight App Spoofing GPS Spoofing Spoofing the Communication link between the drone and the controller
Tampering	Tampering with firmware
Repudiation	Out of research scope
Information Disclosure	Access sensitive information Access to memory data
Denial of service	Jamming the controller Jamming the AR Drone 2.0 GPS
Elevation of privilege	Elevating privilege to the root access after gaining access

2.8 Threats Modeling

This section outlines the key steps and methodologies for categorising the threats vectors found in the context of the AR Drone 2.0. With the information gathered, we use this process to identify and categorise the threats in the AR Drone 2.0 which involves a systematic approach to uncover potential security weaknesses and evaluate their impact on the drone's overall security posture.

STRIDE is widely utilised as a threat simulation technique, making it the preferred choice in this paper's threat modeling approach. The key requirement for the chosen threat modeling approach is the ability to efficiently categorise identified attacks. Consequently, it has been chosen for the literature review (Table 5).

2.9 Summary

The security assessment shows that AR Drone 2.0 is vulnerable to several attacks which include MITM, eavesdropping, unauthorised access, remote access, due to unencrypted wireless connection and no authentication mechanism. The main vulnerability in AR Drone 2.0 is the wireless connection because it is unencrypted. The other vulnerabilities based on network scanning are the FTP, Telnet and UDP because they established unsecure connections to outside world.

The findings revealed specific vulnerabilities in the drone's firmware, software, and network components, highlighting the potential risks and consequences for drone owners and organizations. Recommendations were provided to enhance the security of the AR Drone 2.0, and future research directions were suggested to further advance the field of drone security.

3 Discussion

In this section, we compared the findings and insights from our vulnerability assessment and penetration testing on the AR Drone 2.0 with existing studies and research in the field of UAV security. The purpose of this comparison is to provide an analysis of the vulnerabilities present in the AR Drone 2.0, examining how they align with or differ from other vulnerability types commonly observed in the broader UAV industry.

Additionally, it seeks to compare the attack surface of the AR Drone 2.0 with other UAVs, as discussed in existing UAV security studies. Lastly, it aims to address any unique security challenges and vulnerabilities specific to the AR Drone 2.0 that may not have been extensively covered in existing UAV security studies.

We provided a concise overview of the findings and knowledge acquired from our evaluation of vulnerabilities and penetration testing conducted on the AR Drone 2.0. These findings highlight the distinct importance of our research within the realm of security for Unmanned Aerial Vehicles (UAVs) and commercially available drones, specifically focusing on the AR Drone 2.0.

3.1 Comparison to Existing Research

Existing research on UAV security offers a wide-ranging and thorough perspective on security challenges within the UAV industry. These studies shed light on vulnerabilities, privacy concerns, and regulatory issues, and they provide suggestions for improving UAV security measures. It is imperative to comprehend the insights and patterns derived from these studies to place the security evaluation of the AR Drone 2.0 in context and contribute to the wider conversation surrounding UAV security.

The results of these investigations highlight the necessity for enhanced security measures in the design, development, and operation of UAVs. The studies also emphasise the significance of user education and awareness to diminish security vulnerabilities.

3.2 Comparison of Vulnerability Types

The vulnerabilities identified in the AR Drone 2.0 align with common themes found in existing UAV security studies. These include vulnerabilities related to communication, firmware, and physical security. The weaknesses in authentication and access control mechanisms identified in the AR Drone 2.0 are also indicative of broader concerns surrounding UAV security.

Another recurring concern is privacy, with worries about UAVs capturing sensitive information, infringing on personal privacy, and violating airspace regulations.

The issue of unauthorised access remains a significant challenge within the UAV industry. Concerns relating to communication security vulnerabilities, such as weak encryption and susceptibility to interception, are shared across various UAV models. It is crucial to prioritise the protection of communication channels across the entire spectrum of UAVs. Similarly, firmware vulnerabilities are not exclusive to the AR Drone 2.0 and are frequently reported in existing research. Guaranteeing secure firmware updates and mitigating potential risks associated with code execution are priorities that extend throughout the industry.

3.3 Severity and Impact Comparison

The vulnerabilities found in the AR Drone 2.0, including weak authentication and firmware problems, are consistent with the severity and impact levels commonly observed in previous studies on UAV security. Both the AR Drone 2.0 and other UAVs tend to rate communication vulnerabilities, firmware vulnerabilities, and physical security issues as highly severe and impactful. Privacy implications, which are a significant concern in both the AR Drone 2.0 and UAV security, may differ in severity depending on the sensitivity of the collected data and the applicable legal framework.

3.4 Attack Surface Comparison

The AR Drone 2.0 relies heavily on wireless communication for remote control and data transmission. This wireless communication channel serves as a significant attack surface, potentially susceptible to interception, spoofing, and other wireless attacks.

The attack surface of the AR Drone 2.0, including wireless communication, mobile application, firmware, and sensors, aligns with common attack surfaces identified in existing UAV security studies. Both the AR Drone 2.0 and other UAVs share similar attack surfaces, with wireless communication and software components being particularly critical in terms of vulnerabilities and potential exploitation. Sensors and cameras, while integral to both the AR Drone 2.0 and other UAVs, may have varying levels of sensitivity and exposure depending on the drone's intended use.

3.5 Unique Security Challenges

The AR Drone 2.0 is designed primarily for consumer use, making it more accessible to a broad range of users. This accessibility can lead to security challenges as it may be operated by individuals with varying levels of technical expertise and awareness of security best practices. The AR Drone 2.0 relies on a mobile app for control and configuration. Mobile apps can introduce unique security challenges, including potential vulnerabilities in the app itself, as well as security risks associated with app permissions and user behaviour.

Being a consumer-grade drone, the AR Drone 2.0 may have legacy systems or outdated components that are no longer actively maintained or patched by the manufacturer. These legacy elements can pose security risks if vulnerabilities exist and cannot be addressed through updates.

Consumer drone users may not always be aware of the security risks associated with their devices. This lack of awareness can lead to unsafe practices, such as flying drones in restricted airspace or failing to secure their home Wi-Fi networks, which can be used to control the drone.

Consumer drones like the AR Drone 2.0 are often used for recreational photography and videography. This raises privacy concerns, as drones can inadvertently capture images and videos of individuals without their consent, potentially violating privacy regulations.

3.6 Contributions and Insights

Our research has identified and documented specific vulnerabilities in the AR Drone 2.0's security posture. These vulnerabilities include weaknesses in authentication mechanisms, communication security, firmware, and physical access controls. We conducted a thorough security assessment that encompassed multiple facets of UAV security, including authentication, communication, firmware analysis, and physical security. This comprehensive approach provides a holistic view of the AR Drone 2.0's vulnerabilities. Our research offers practical and actionable recommendations for mitigating the identified vulnerabilities. These recommendations cover areas such as firmware updates, communication encryption, authentication improvements, and physical security measures.

We have compared the vulnerabilities and security challenges identified in the AR Drone 2.0 with existing UAV security studies. This comparison highlights commonalities and distinctions, contributing to a better understanding of security trends in the UAV industry. Our research provides insights into the unique security challenges associated with consumer-grade drones like the AR Drone 2.0. These insights include considerations related to mobile app integration, community involvement, legacy components, limited user security awareness, and privacy concerns.

Our research confirms that the AR Drone 2.0 shares common security challenges with other UAVs, such as vulnerabilities in communication, firmware, and physical access controls. This underscores the importance of addressing well-documented security issues in consumer-grade drones.

The research highlights the significance of user awareness and responsible practices in UAV security. Consumer drone users may not always be fully aware of the security implications of their actions, emphasizing the need for educational efforts.

Privacy concerns related to drones, especially those used for recreational photography and videography, need to be addressed. Capturing images and videos without consent can lead to privacy violations and regulatory issues. The involvement of drone enthusiasts and communities introduces both opportunities and risks. While innovation can thrive, community activities must be conducted responsibly to avoid security breaches.

Our research on the AR Drone 2.0 makes significant contributions by identifying specific vulnerabilities, offering actionable recommendations for mitigation, comparing our findings with existing UAV security studies, and shedding light on unique security challenges in the consumer drone market. These contributions and insights contribute to the ongoing discourse on UAV security and can guide efforts to enhance the security of consumer-grade drones like the AR Drone 2.0.

3.7 Recommendations

The security enhancements provided in this research are based on vulnerabilities assessment and penetration testing carried out on AR Drone 2.0.

- **Secure Network Configuration**
  The communication channel is unencrypted, which allows any user to connect without a password. This is a vulnerability which has a greater impact on the device. This can lead to eavesdropping and man-in-the-middle attack. Drone operators can protect critical data and stop unauthorised access or theft by utilising encryption. Encryption technique between the drone and the ground station or the remote controller can be employed.
  When choosing an encryption method, it is however essential to choose a strong encryption mechanism because weak encryptions are also susceptible to attacks. Implementation of strong encryption (such as WPA2 or WPA3) in a layered protocol is one of the best ways to mitigate security risk and unauthorised access.
- **Secure Communication Channels**
  To mitigate the risk of Man-in-the-Middle attacks, it is recommended to establish secure communication channels between the AR Drone 2.0 and its controller. This can be achieved by implementing strong encryption protocols, such as Transport Layer Security (TLS), to ensure the confidentiality and integrity of the communication.
- **Secure use of FTP and Telnet**
  Protecting the wireless connection automatically gives room for protection over FTP and Telnet because there won't be unauthorised connection to the outside world. However, if a brute-force attack is successful, a malicious actor might have a full privilege access to the drone services and operating system through the FTP and Telnet services. Thus, a secure operation is recommended for both services. In order to protect against unauthorised access and malicious behaviour, authentication can be employed to ensure that only authorised users have access to drone control systems.

Migrating to a secure shell protocol provides features to protect the confidentiality, integrity, and authentication. This will protect against malicious upload or download of sensitive and confidential data.

- **Ports closing**

  Drones always remain operational, keeping its services active and associated ports accessible. For instance, when the drone is in flight, the FTP and telnets ports are exposed to potential security threats.

  To partially address this vulnerability and enhance drone security, one approach is to deactivate the services when they are not required. This can be accomplished by either terminating the services or implementing inbound port blocking. Iptables, a firewall utility, can be utilized to discard all packets intended for unused services, while permitting legitimate packets during authorized usage.
- **Redundancy components**

  Only the security element of availability is crucial for the gyroscope and accelerometer components. Redundancy of components is required for these sensors to assure their availability even in the event of a breakdown.

  Sensitive data should be separated from system data to ensure integrity. The ROM and other associated resources should be partitioned.

  Applications running on the Drone must not be allowed to have access to each other's data. This will provide some certain level of confidentiality to user's data. Data such as live video streaming should be encrypted to protect it against eavesdropping and MITM.
- **Firmware Updates and Patch Management**

  One of the primary recommendations is to ensure regular firmware updates and patch management for the AR Drone 2.0. Manufacturers should release timely security updates to address known vulnerabilities and security weaknesses. Users should be encouraged to regularly update the firmware of their drones to ensure they are protected against the latest threats.
- **Awareness and Training**

  Both the drone pilots and operators should have proper training and education about cyber security risk related to drones as well as how to recognise and respond to potential dangers.
- **Collaboration and information sharing**

  Collaboration between the manufacturers and the industries that use drone is important to ensure effective security measures. Sharing information such as threat intelligence and best practices between manufacturers, operators and researchers is also an important effort to address and contribute to drone cybersecurity challenges.

4 Experiment Results

AR.FreeFlight 2.4.15 application was installed on the Apple iPhone 6s. Numerous test flights were carried out to take pictures and make videos to collect and analyse data for the experiment. The above-mentioned smartphone was used in the scenarios to control the drone to capture data while the windows laptop was used to carry out the exploits. Based on the results of the vulnerabilities assessment, attacks were launched to discover relevant evidence and determine the useful and relevant information stored in the drone, flight patterns, and stored media, including photographs, videos, and logs.

The communication channel between the iOS and the drone is unencrypted, which allows any user to connect without a password. This is a vulnerability which has a greater impact on the device. This can lead to eavesdropping and man-in-the-middle attack. Cisco network adapter was put in monitor mode to eavesdrop the communication between AR Drone 2.0 and the controller, also, eavesdrop on the live streaming video show that AR Drone 2.0.

After connecting to the network through Iphone 6s (AR.FreeFlight 2.4.15 software installed on it), which serves as the controller for controlling the Parrot AR Drone 2.0, the Ip address of the drone was revealed.

Below picture is the interface of the AR.FreeFlight 2.4.15 software for AR Drone 2.0. It has connection information, the battery, photo and video icon, altitude information and so on. The AR.Freeflight APP stores the personal identifiable information, flight data logs (Accelerometer, battery level, ultrasonic altimeter), captured media (photos and videos) (Fig. 4).

Fig. 4 FreeFlight 2.4.15 application interface

```
┌──(k99㉿kali)-[/]
└─$ sudo nmap -sS -A -T4 -p- 192.168.1.1 -Pn
[sudo] password for k99:
Starting Nmap 7.92 ( https://nmap.org ) at 2023-07-14 06:42 CDT
Stats: 0:10:26 elapsed; 0 hosts completed (1 up), 1 undergoing Script Scan
NSE Timing: About 98.21% done; ETC: 06:52 (0:00:07 remaining)
Stats: 0:10:57 elapsed; 0 hosts completed (1 up), 1 undergoing Script Scan
NSE Timing: About 98.21% done; ETC: 06:53 (0:00:08 remaining)
Nmap scan report for 192.168.1.1
Host is up (0.0068s latency).
Not shown: 65528 closed tcp ports (reset)
PORT     STATE SERVICE        VERSION
21/tcp   open  ftp            BusyBox ftpd (D-Link DCS-932L IP-Cam camera)
| ftp-anon: Anonymous FTP login allowed (FTP code 230)
| drwxr-xr-x    2 0        0             160 Jan  1  2000 boxes
|_-rw-r--r--    1 0        0           48186 Jan  1  2000 police-notice.html.gz
| ftp-syst:
|   STAT:
| Server status:
|  TYPE: BINARY
|_Ok
|_ftp-bounce: bounce working!
23/tcp   open  telnet
| fingerprint-strings:
|   GenericLines, NULL:
|     BusyBox v1.14.0 () built-in shell (ash)
|     Enter 'help' for a list of built-in commands.
|   GetRequest:
|     HTTP/1.0
|     BusyBox v1.14.0 () built-in shell (ash)
```

Fig. 5 Full aggressive NMAP scan

Nmap scan shows the open ports and the services with the versions of the service running on each port. Also, it reveals some of the documents in each of the ports and the login information. As for the ftp, no authentication was required to login, although "Anonymous" can be used to login to the ftp. Telnet does not require login authentication as well. Furthermore, we used Wireshark to capture live stream video packet and communication traffic between the drone and the controller (Figs. 5, 6).

No authentication was required to gain access to both FTP and Telnet. After gaining access to the ftp, we were able to upload files through the FTP.

After gaining access to the telnet, UID syntax was ran and it was discovered that we have "Root" access. Root access basically allows users to perform any task on the drone (Figs. 7, 8, 9).

We navigated to the "bin" directory, and below pictures shows the list of commands that can be performed after gaining a root access to the AR Drone 2.0 system. Some of the commands listed below are called "Destructive commands" which when used by malicious actor, can cause harm or injuries to people (Fig. 10).

Firmware directory shows the "GPS" configuration files. These files can be modified to suit bad actors' preferences or redirect the AR Drone 2.0 to another destination (Figs. 11, 12, 13, 14, 15, 16).

"Reboot" command was issued while the drone was in operation. The following picture shows the result after the command was issued. The connection link between

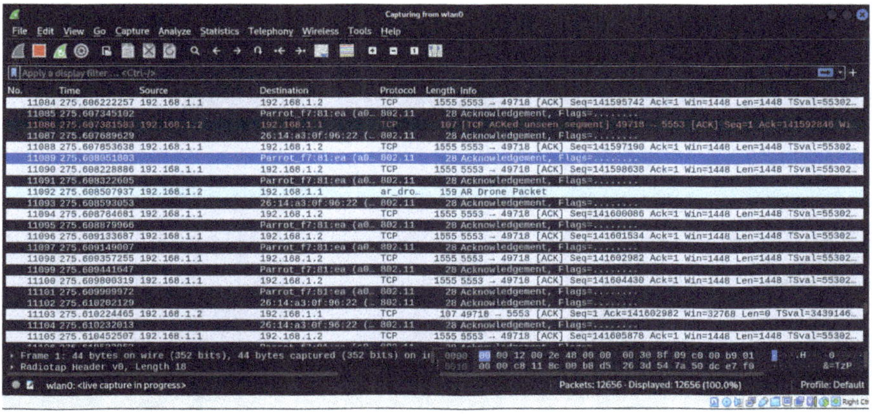

Fig. 6 Live video analysis/monitoring using Wireshark

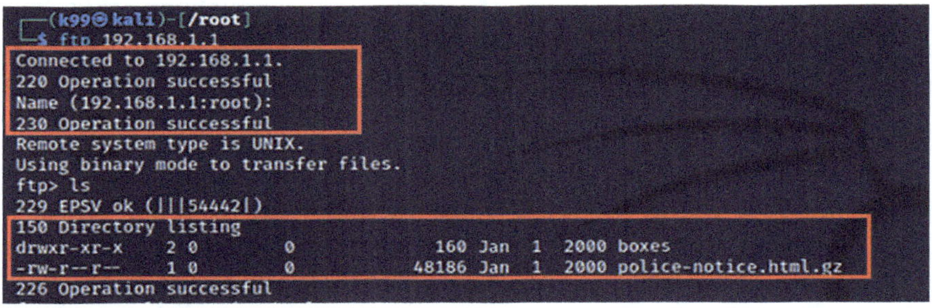

Fig. 7 FTP no authentication login

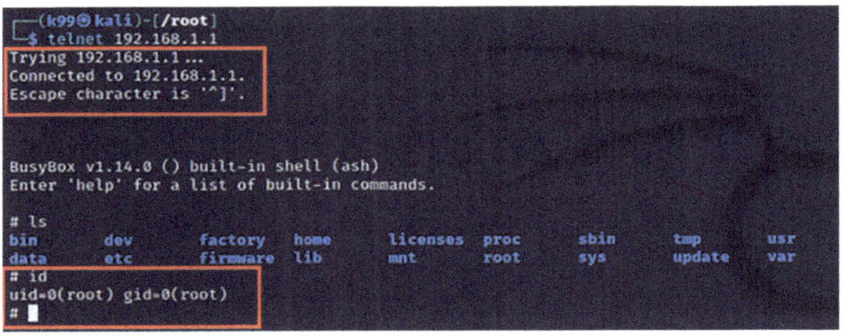

Fig. 8 Telnet no authentication login

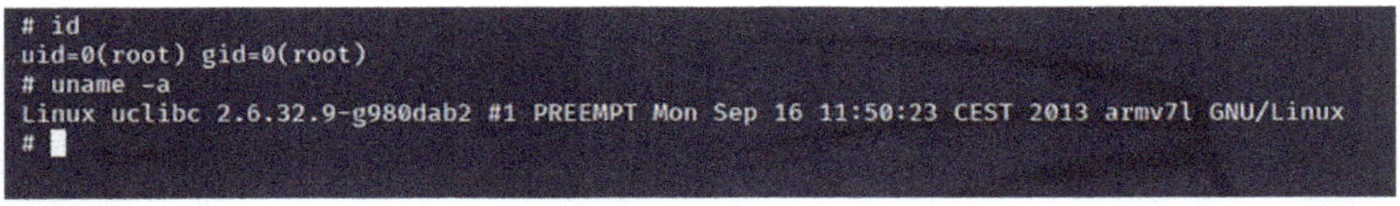

Fig. 9 UID syntax

Fig. 10 /bin directory

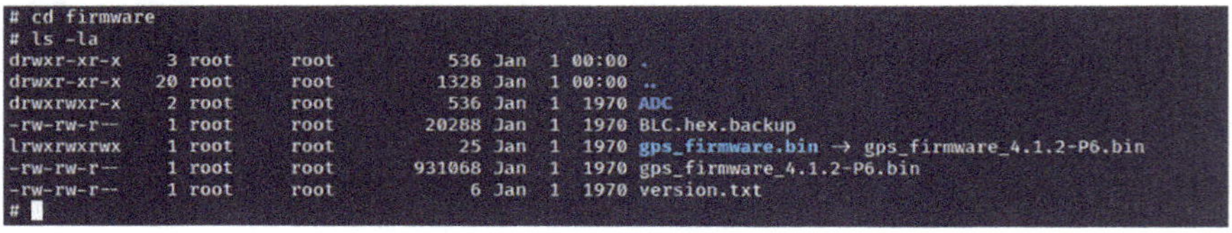

Fig. 11 /firmware directory

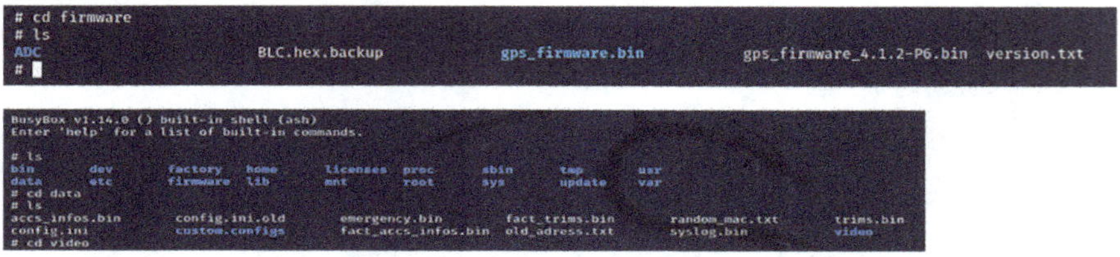

Fig. 12 /data directory

the controller and the drone was cut-off and the authorised user was unable to control the AR Drone anymore (Figs. 17, 18).

Nessus vulnerability assessment revealed which vulnerabilities are critical and which are medium. No matter what the scale is, it is all vulnerability which are harmful and can be exploited by bad actors (Figs. 19, 20, 21, 22, 23).

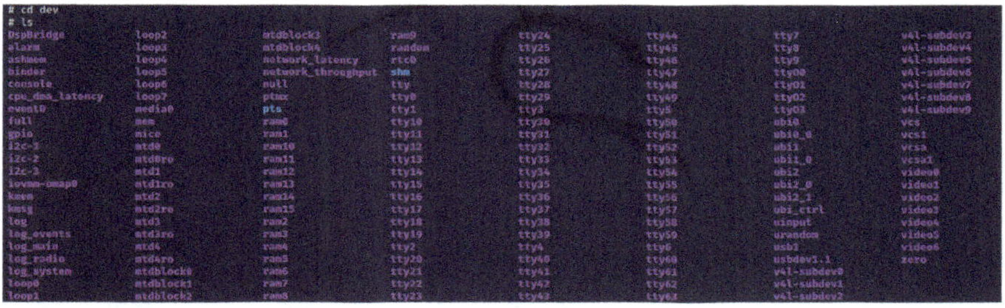

Fig. 13 /dev directory

Fig. 14 /factory directory

Fig. 15 /sys directory

5 Conclusion and Future Research

This research highlights the vulnerabilities inherent in the AR Drone 2.0 and the significance of conducting penetration testing assessments to ensure the security of unmanned aerial vehicles. By understanding and addressing these vulnerabilities, drone operators can enhance the overall security posture of the AR Drone 2.0 and protect against potential threats.

The existing vulnerabilities and attack vectors associated with drone security pose a significant threat to the security of drone systems. Threat actors can exploit these vulnerabilities and attack vectors to compromise the security of drone systems and cause significant damage. It is important for drone operators to be aware of these vulnerabilities and attack vectors and take measures to mitigate them.

The research answers the research questions by providing a thorough analysis of vulnerabilities and conducting a penetration testing assessment on the AR Drone 2.0. Section 4 discussed more about this. It can be deduced that Parrot AR Drone 2.0 has no security measures in place both in design, communication, and software.

Ultimately, safeguarding the AR Drone 2.0 against vulnerabilities requires a multi-layered approach that combines proactive measures with continuous security awareness and timely implementation of security updates. By adopting these practices,

```
# ls -la /sbin
drwxrwxr-x    2 root     root          2752 Jan  1  1970 .
drwxr-xr-x   20 root     root          1328 Jan  1 00:00 ..
lrwxrwxrwx    1 root     root            14 Jan  1  1970 arp → ../bin/busybox
lrwxrwxrwx    1 root     root            14 Jan  1  1970 blkid → ../bin/busybox
-rwxr-xr-x    1 root     root         17192 Jan  1  1970 blkid-ng
lrwxrwxrwx    1 root     root            14 Jan  1  1970 devmem → ../bin/busybox
lrwxrwxrwx    1 root     root            14 Jan  1  1970 fbsplash → ../bin/busybox
lrwxrwxrwx    1 root     root            14 Jan  1  1970 fdisk → ../bin/busybox
lrwxrwxrwx    1 root     root            14 Jan  1  1970 halt → ../bin/busybox
lrwxrwxrwx    1 root     root            14 Jan  1  1970 hwclock → ../bin/busybox
lrwxrwxrwx    1 root     root            14 Jan  1  1970 ifconfig → ../bin/busybox
-rwxr-xr-x    1 root     root         20096 Jan  1  1970 ifrename
lrwxrwxrwx    1 root     root            14 Jan  1  1970 init → ../bin/busybox
lrwxrwxrwx    1 root     root            14 Jan  1  1970 insmod → ../bin/busybox
-rwxr-xr-x    1 root     root         20204 Jan  1  1970 iwconfig
-rwxr-xr-x    1 root     root         11408 Jan  1  1970 iwevent
-rwxr-xr-x    1 root     root          7840 Jan  1  1970 iwgetid
-rwxr-xr-x    1 root     root         23852 Jan  1  1970 iwlist
-rwxr-xr-x    1 root     root         10292 Jan  1  1970 iwpriv
-rwxr-xr-x    1 root     root          7852 Jan  1  1970 iwspy
lrwxrwxrwx    1 root     root            14 Jan  1  1970 klogd → ../bin/busybox
lrwxrwxrwx    1 root     root            14 Jan  1  1970 lsmod → ../bin/busybox
lrwxrwxrwx    1 root     root            14 Jan  1  1970 mdev → ../bin/busybox
lrwxrwxrwx    1 root     root            14 Jan  1  1970 mkdosfs → ../bin/busybox
lrwxrwxrwx    1 root     root            14 Jan  1  1970 mkfs.vfat → ../bin/busybox
lrwxrwxrwx    1 root     root            14 Jan  1  1970 modprobe → ../bin/busybox
-rwxr-xr-x    1 root     root          7248 Jan  1  1970 mqueue
lrwxrwxrwx    1 root     root            14 Jan  1  1970 poweroff → ../bin/busybox
lrwxrwxrwx    1 root     root            14 Jan  1  1970 reboot → ../bin/busybox
lrwxrwxrwx    1 root     root            14 Jan  1  1970 rmmod → ../bin/busybox
lrwxrwxrwx    1 root     root            14 Jan  1  1970 route → ../bin/busybox
lrwxrwxrwx    1 root     root            14 Jan  1  1970 setconsole → ../bin/busybox
lrwxrwxrwx    1 root     root            14 Jan  1  1970 switch_root → ../bin/busybox
```

Fig. 16 Destructive commands listed in "sbin" directory

```
# reboot
# Connection closed by foreign host.
```

Fig. 17 Reboot command

Fig. 18 Disconnection from main/original controller

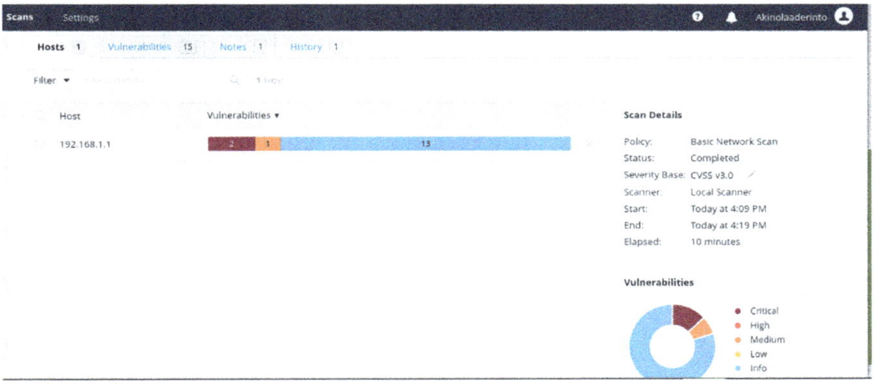

Fig. 19 Nessus scan details

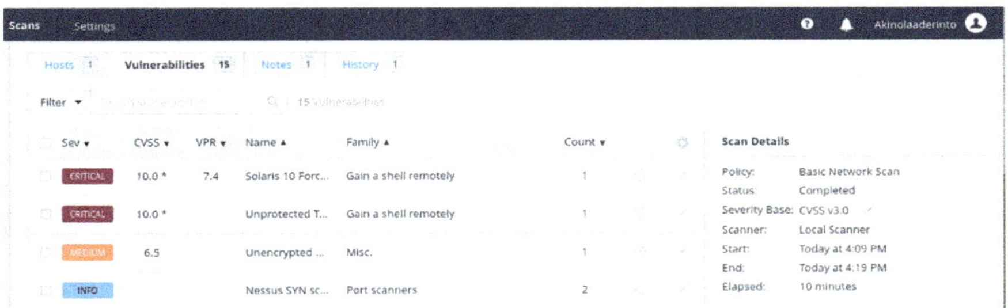

Fig. 20 Nessus scan vulnerabilities result

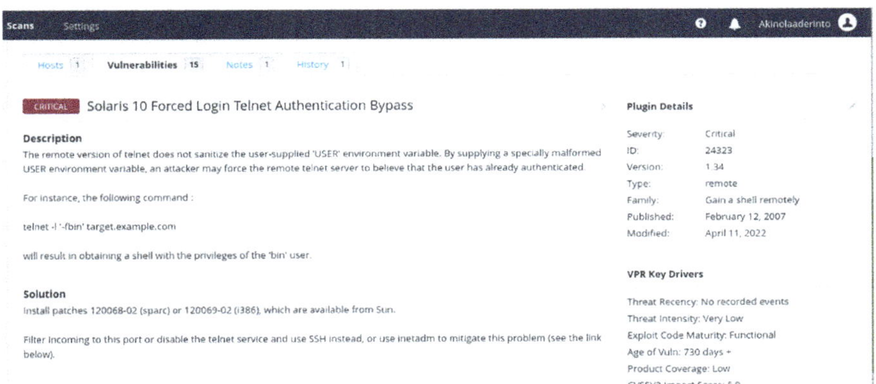

Fig. 21 Nessus scan vulnerabilities result details

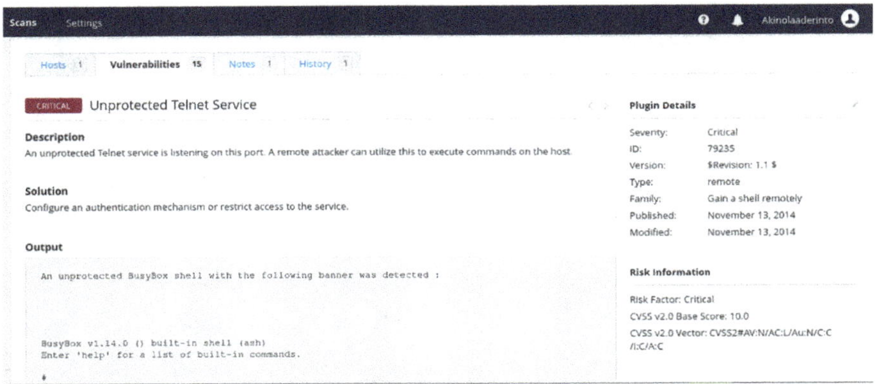

Fig. 22 Nessus scan vulnerabilities result details

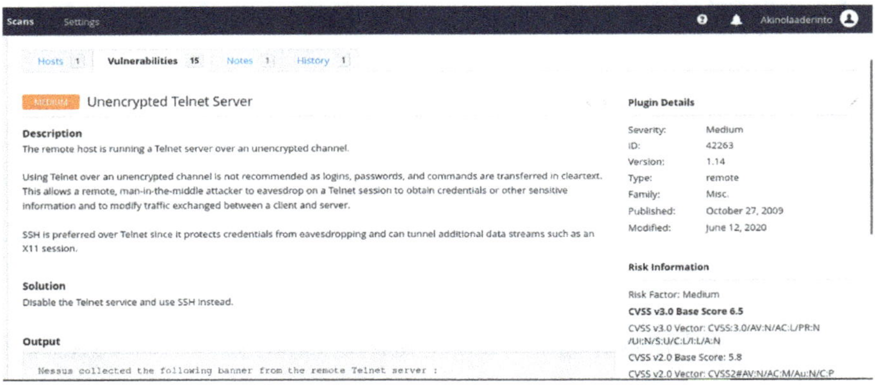

Fig. 23 Nessus scan vulnerabilities result details

drone operators can ensure the safe and secure operation of the AR Drone 2.0, protecting both their assets and the privacy of individuals and organizations involved.

However, it is important to note that there is no single solution that can fully address all the vulnerabilities and attack vectors. A combination of different solutions, such as encryption, authentication, and anti-spoofing techniques, may be necessary to ensure the security of drone systems.

It is important for future research to delve deeper into specific vulnerabilities and develop advanced techniques for identifying and mitigating risks associated with the AR Drone 2.0. Future work can focus on further enhancing the penetration testing framework for the AR Drone 2.0. This can include expanding the scope of vulnerabilities considered, incorporating more advanced attack techniques, and integrating threat intelligence to identify emerging threats.

References

Amrami, D. (2023). *Drones: Security, Vulnerability, Exploitation, Awareness, and Mitigation*. LinkedIn. https://www.linkedin.com/pulse/drones-security-vulnerability-exploitation-awareness-dror-amrami/

Astaburuaga, I., Lombardi, A., La Torre, B., Hughes, C., & Sengupta, S. (2019). Vulnerability Analysis of Ar.Drone 2.0, an Embedded Linux System. In *2019 IEEE 9th Annual Computing and Communication Workshop and Conference (CCWC)*. https://doi.org/10.1109/ccwc.2019.8666464

Bouafif, H., Kamoun, F., Iqbal, F., & Marrington, A. (2018). Drone Forensics: Challenges and New Insights. In *2018 9th IFIP International Conference on New Technologies, Mobility and Security (NTMS)*. https://doi.org/10.1109/ntms.2018.8328747

Choudhary, G., Sharma, V., Gupta, T., Kim, J., & You, I. (2018). *Internet of Drones (IoD): Threats, Vulnerability, and Security Perspectives*. arXiv preprint arXiv:1808.00203

Clark, R. A., Punzo, G., MacLeod, C. N., Dobie, G., Summan, R., Bolton, G., Pierce, S. G., & Macdonald, M. (2017). Autonomous and Scalable Control for Remote Inspection with Multiple Aerial Vehicles. *Robotics and Autonomous Systems, 87*, 258–268.

Hamza, P., & Qassim, A. (2021). *How Secure Are Drones? A Security Analysis of a Drone System.*

Hamdi, D. A., Iqbal, F., Alam, S., Kazim, A., & MacDermott, A. (2019). Drone Forensics: A Case Study on DJI Phantom 4. In *2019 IEEE/ACS 16th International Conference on Computer Systems and Applications (AICCSA)*. https://doi.org/10.1109/aiccsa47632.2019.9035302

Iqbal, F., Alam, S., Kazim, A., & MacDermott, Á. (2019). Drone Forensics: A Case Study on DJI Phantom 4. In *2019 IEEE/ACS 16th International Conference on Computer Systems and Applications (AICCSA)* (pp. 1–6). IEEE.

Kharchenko, V., & Torianyk, V. (2018, May). Cybersecurity of the Internet of Drones: Vulnerabilities Analysis and Imeca based Assessment. In *2018 IEEE 9th International Conference on Dependable Systems, Services and Technologies (DESSERT)* (pp. 364–369). IEEE.

Kumar, C. R. S., & Mohanty, S. (2021). Current Trends in Cyber Security for Drones. In *2021 International Carnahan Conference on Security Technology (ICCST)*. https://doi.org/10.1109/iccst49569.2021.9717376

Marty, J. A. (2014). *Vulnerability Analysis of the Mavlink Protocol for Command and Control of Unmanned Aircraft.*

Shafique, A., Mehmood, A., & Elhadef, M. (2021). Survey of Security Protocols and Vulnerabilities in Unmanned Aerial Vehicles. *IEEE Access, 9*, 46927–46948. https://doi.org/10.1109/access.2021.3066778

Siddiqi, M. A., Iwendi, C., Jaroslava, K., & Anumbe, N. (2022). Analysis on Security-Related Concerns of Unmanned Aerial Vehicle: Attacks, Limitations, and Recommendations. *Mathematical Biosciences and Engineering, 19*(3), 2641–2670. https://doi.org/10.3934/mbe.2022121

Stanković, M., Mirza, M. M., & Karabiyik, U. (2021). UAV Forensics: DJI Mini 2 Case Study. *Drones, 5*(2), 49. https://doi.org/10.3390/drones5020049

Yaacoub, J.-P., Noura, H., Salman, O., & Chehab, A. (2020a). Security Analysis of Drones Systems: Attacks, Limitations, and Recommendations. *Internet of Things, 11*, 100218. https://doi.org/10.1016/j.iot.2020.100218

Yaacoub, J.-P.A., Noura, M., Noura, H. N., Salman, O., Yaacoub, E., Couturier, R., & Chehab, A. (2020b). Securing Internet of Medical Things Systems: Limitations, Issues and Recommendations. *Future Generation Computer Systems, 105*, 581–606. https://doi.org/10.1016/j.future.2019.12.028

Quantum Technologies for Space and Aerial Vehicles

Viduranga Shenal Landers

Abstract Quantum computing has the potential to revolutionize numerous fields, including satellite systems and unmanned aerial vehicles (UAVs). This chapter introduces quantum computing and discusses how it could be used to address major challenges in satellite and UAV technologies, such as navigation, control, data privacy, and secure communication, which have emerged as a result of the growing use of satellites and UAVs. Quantum technologies have the potential to facilitate secure communication, enhanced forensics, and substantial accelerations in computations such as path optimization, searching and machine learning. As we approach the quantum era, Quantum-resistant encryption protocols and Quantum Key Distribution (QKD) protocols will be crucial future countermeasures against cyber threats aimed at satellites and UAVs. In addition to these applications, governance and standardization of quantum technologies, as well as their use in satellite systems and unmanned aerial vehicles, are discussed.

Keywords Quantum computing · QKD · Space · UAV · Policing

1 Introduction

Quantum technologies have become a hot topic in scientific and technological circles, with its potential to revolutionize numerous industries and redefine our understanding of technological possibilities. Among them, quantum computing, quantum communication, and quantum sensing stand out as particularly promising for a range of industries (Sepulveda, 2022). Quantum computers, based on the strange principles of quantum mechanics, can perform complex computations far beyond the reach of even the most powerful classical computers (Arute et al., 2019). Classical sensing has sufficed for many practical scenarios, but quantum sensing takes precision to a whole

V. S. Landers (✉)
University of Colombo School of Computing, Colombo, Sri Lanka
e-mail: vidurangalanders@gmail.com

Space Generation Advisory Council, Vienna, Austria

H. Jahankhani et al. (eds.), *Space Governance*, Space Law and Policy,
https://doi.org/10.1007/978-3-031-62228-1_4

new level, offering exciting possibilities for scientific research, navigation, and more (Bongs et al., 2023). Quantum communication, leveraging quantum states of particles, promises unbreakable security for information transmission (Gisin & Thew, 2007). While the promises have been verified in theory, practical implementations are still in their early stages of development.

An emerging field where quantum technologies holds immense potential is the domain of space exploration and aviation. With the advent of fully autonomous systems, aerial vehicles are now capable of performing a wide range of tasks and are becoming an integral part of day to day life (Elmokadem & Savkin, 2021). These Unmanned Aerial Vehicles (UAVs) now execute a range of operations such as aerial surveillance (Engberts & Gillissen, 2016), disaster response (Estrada & Ndoma, 2019; Marzena Półka, 2017), package delivery (Applin, 2016) and military purposes (Lee, 2016). Recent advancements in small satellites, autonomous rovers, and increased accessibility to space has given access to a wealth of new realms with potential commercial value (Bushnell, 2021; Spremo et al., 2019). To fully exploit these possibilities, advanced autonomous systems with robust data analysis capabilities are essential.

Current state-of-the-art navigation, communication, and data processing systems within satellites and UAVs face two major challenges: computational limitations and increasing cybersecurity threats. Quantum technology has the transformative power to push the limits of efficiency, accuracy, and security in space and aerial vehicles. Applications such as quantum-enhanced navigation systems, unbreakable satellite communication networks, and accelerated on-board computations hold the key to unlocking new possibilities. Quantum technologies have the potential to not only enhance existing computational capabilities but also act as a robust countermeasure against evolving cyber threats (IBM Institute for Business Value, 2023). Addressing the emerging cyber threats of the quantum era requires the development of quantum-resistant encryption and ongoing efforts to protect sensitive data collected and transmitted by satellites and UAVs.

Technological advancements must be accompanied by responsible governance and standardization. This includes critically engaging with emerging ethical problems and fostering international collaboration to ensure the safe, responsible, and equitable deployment of quantum technologies in space and aerial domains.

In this chapter, we explore the profound impact of quantum technology on space and aerial vehicles. We begin by examining the advancements in the aerospace domain and the challenges posed by the limitations of current computing methods (Sect. 2). We then discuss how quantum computing can potentially solve these problems, drawing on examples from existing literature (Sect. 3). Next, we delve into the realm of quantum communication, with a particular focus on Quantum Key Distribution (QKD). We highlight the potential of QKD to revolutionize secure communication in space and air (Sect. 4). However, we also acknowledge the challenges faced by current quantum systems (Sect. 5). Finally, we examine the critical aspects of governance and standardization for quantum technologies. We explore the need for a robust regulatory framework and the importance of developing common standards to ensure the responsible and effective use of quantum technology in the aerospace domain (Sect. 6).

2 Quantum Computing Applications for Spacecraft and UAVs

In the earlier decades, humans were just starting to explore the benefits of space and aerial travel. Given the high risks associated with space and air travel, it is not a mystery how the concept of unmanned vehicles came into being. Back then, humans had the luxury of being the most advanced beings with the highest intellectual capability and controlled autonomy. Therefore, more focus was given to building systems that could ensure the safety of a human operator rather than fully automated machinery. However, as systems evolved and became more complex, the need for automation increased. As technologies made sufficient advancements, automation too became advanced to the level of fully autonomous systems (Little, 2019).

In the modern era, with the exponential advancements in Artificial Intelligence (AI) and robotics, the landscape of space exploration and aviation is rapidly shifting. At first, automation was focused towards reducing human operator responsibilities, but with the complete automation of aerial vehicles, different applications of aerial systems in day to day activities have emerged (Elmokadem & Savkin, 2021). While the realm of space has favored unmanned systems, recent advancements in small satellites, autonomous rovers, and low-cost launch facilities has given access to a wealth of information and data sources to extend humanity's understanding of the cosmos (Bushnell, 2021). This shift towards increasingly intelligent and unmanned systems offers the potential for increased efficiency, reduced risk to human life, and access to regions otherwise inaccessible.

Despite these remarkable advancements in autonomous systems, challenges regarding navigation, communication, data processing, and security persist. Therefore, in this section, we will look at the different challenges that aerospace systems face that could be addressed with quantum computing.

Navigation and Control: As space and aerial systems become increasingly miniaturized, the use of swarms and constellations is rapidly growing. Coordinating and controlling these large-scale systems in real-time poses significant challenges, requiring efficient scheduling and optimization algorithms (Abdelkader et al., 2021). Further, trajectory optimization for vehicles operating in dynamic environments faces similar computational bottlenecks. One such example is calculating trajectories for multi-body problems (Earth-Moon-Spacecraft) or those with gravitational assist maneuvers. Another similarly hard problem class is the computations related to the Travelling Salesman Problem (Matai et al., 2010). Classical computers often struggle to find optimal solutions for such problems in a timely manner, leading to inefficiencies and delays.

Data processing and Analytics: Aerial and space missions often generate enormous amounts of data, from high-resolution imagery to environmental sensor readings. Classical computers on board these systems have difficulty handling the complex demands of analyzing such data in real-time, which can lead to missed opportunities or delayed decision-making. Furthermore, extracting meaningful features and insights from large-scale datasets is a challenging task. Classical machine learning (ML) algorithms often require extensive training and can be computationally expensive, making them impractical for real-time applications (Mohsan et al., 2023).

Communication: Conventional communication protocols and encryption schemes used in both satellites and UAVs were not designed with the threat of quantum computers in mind. These protocols rely on mathematical problems that are nearly impossible for classical computers to solve. As quantum computing advances, these systems face potential vulnerabilities, making them susceptible to eavesdropping and data breaches (Wolf, 2021). Ensuring the security of these systems against unauthorized access, data theft, and denial-of-service attacks is a major challenge.

Addressing all of these interrelated challenges is crucial to pave the way for secure, reliable, and truly versatile space and aerial vehicles. Most of the above problems exist due to the limitations of current classical systems. Can quantum computing address these challenges?

2.1 Quantum Computing

Quantum computing represents a significant advancement in information processing. However, to assess its potential for solving the challenges in aerospace systems, we must first grasp its fundamental principles. Quantum computing leverages principles of quantum mechanics such as superposition, entanglement, and interference for computation and information processing (Nielsen & Chuang, 2000).

Unlike classical computers, which rely on binary bits representing either a 0 or 1, quantum computers harness the strangeness of quantum mechanics. Within this realm, quantum bits, or "qubits", can exist as both a 0 and 1 simultaneously—a phenomenon known as superposition. Quantum computers also makes use of another fascinating quantum property called entanglement. This principle describes how qubits can become inextricably linked, such that manipulating one immediately affects the other, even at large distances. Quantum computers utilize quantum interference by manipulating the probabilistic waves associated with qubits through carefully designed sequences of operations which allows them to steer calculations toward the correct answer. There are two prominent methods in practical quantum computing: gate model and quantum annealing. In the gate model, the sequences of operations are known as quantum circuits and consists of quantum gates that acts upon qubits (McGeoch et al., 2019). These quantum principles allow quantum computers to tackle problems with a vast number of possible solutions, significantly outperforming classical computing in specific use cases (Nielsen & Chuang, 2000).

One of the important achievements in the field of quantum computing is the Shor's algorithm (Shor, 1997). It has the theoretical potential to undermine commonly used encryption schemes by efficiently factoring large numbers. Another similar achievement is the Grover's algorithm, which promises a quadratic speedup over traditional methods when searching unstructured databases, a capability that could revolutionize fields heavily dependent on rapid search capability (Grover, 1996).

While certain quantum algorithms hold the promise of surpassing the capabilities of classical algorithms, it is crucial to acknowledge that such algorithms often require ideal quantum computers with a large number of high-quality qubits. Even though ideal quantum computers are far from reality, these novel algorithms present many new opportunities. The current state of quantum computing is known as the NISQ era, which stands for Noisy Intermediate Scale Quantum (Preskill, 2018). Characterized by a limited number of qubits, high error rates, and short coherence times, NISQ devices pose hurdles for running complex quantum algorithms.

Despite these constraints, the NISQ era has witnessed remarkable progress. Quantum algorithms are not merely theoretical concepts; they have been successfully implemented on noisy quantum devices. Shor's factorization algorithm, for instance, has demonstrated its abilities in factoring small numbers on real quantum hardware back in 2001 (Vandersypen et al., 2001). Researchers continue to make strides in improving quantum hardware, reducing error rates, and developing novel quantum algorithms. In addition, there are a number of companies and startups working to develop quantum software and applications for real-world applications on NISQ devices.

While the NISQ era is still in its infancy, the potential applications of quantum computing are vast. From tackling intricate optimization problems (Qiskit Optimization Development Team, 2024) to accelerating machine learning algorithms (Schuld & Petruccione, 2021) and simulating quantum systems (Johnson et al., 2014), quantum computers hold promise for revolutionizing fields such as finance, logistics, drug discovery, natural language processing, computer vision, medical diagnosis, and materials science.

Before exploring the potential applications of quantum computers in the aerospace sector, it is crucial to understand their advantages over classical computers. These advantages can be realized in two ways. Firstly, one can directly examine well-researched applications. Secondly, it is essential to grasp the general capabilities of quantum computers and identify potential application areas with empirical evidence for similarities.

Quantum computers are known to excel at three types of problems: optimization problems, processing large data volumes, and simulating quantum systems. Quantum key distribution, though often categorized as a quantum computing technology, will be discussed separately in this chapter under quantum communication.

2.2 Solving Optimization Problems

Quantum optimization algorithms presents a potential for exponential speedup in solving complex optimization problems compared to classical algorithms. The power of quantum optimization algorithms lies in harnessing the unique properties of quantum mechanics to expand the computational dimensions by exploring number of solutions in parallel. This parallelism enables quantum optimization algorithms to efficiently navigate the complexities of optimization problems, significantly reducing the computational time required to find optimal or near-optimal solutions.

One prominent example of a quantum optimization algorithm is the Quantum Approximate Optimization Algorithm (QAOA), which combines classical and quantum computation to tackle optimization problems. QAOA leverages quantum systems to construct a candidate solution and then iteratively refines it using classical optimization techniques (Farhi et al., 2014) (Fig. 1).

Quantum optimization algorithms hold immense promise across various industries and applications. In finance, they can optimize complex financial portfolios, reducing risk and maximizing returns. In healthcare, they can facilitate the discovery of new drugs and therapies by optimizing molecular structures. In logistics, they can optimize supply chain networks, enhancing efficiency and reducing costs.

Despite their potential, quantum optimization algorithms are still in their early stages of development. Several challenges need to be addressed, such as the scalability of quantum systems, the development of efficient quantum algorithms, and the mitigation of errors inherent in quantum computation.

However, advancements in quantum annealing has provided the ability to solve real world optimization problems on NISQ devices known as quantum annealers. Quantum annealing is a quantum optimization algorithm inspired by the physical process of annealing. It uses quantum fluctuations to find the global minimum of a given objective function (McGeoch et al., 2019).

As quantum computing technology continues to advance, quantum optimization algorithms are poised to revolutionize the way we approach and solve complex optimization problems. By combining the power of quantum mechanics with classical

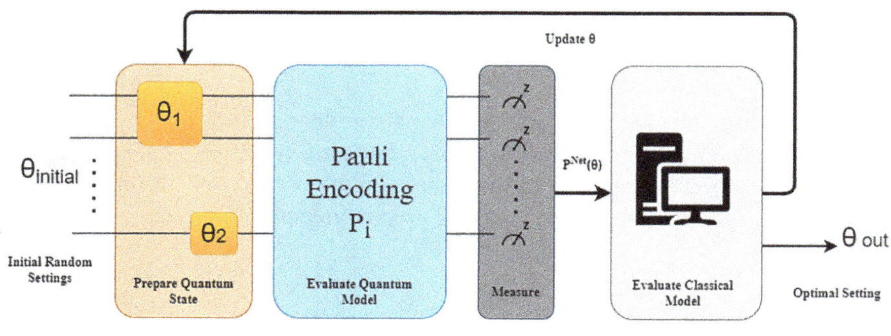

Fig. 1 Variational quantum optimization

computational techniques, quantum optimization algorithms offer unprecedented opportunities for organizations to unlock new frontiers of innovation and achieve breakthroughs in various domains.

2.3 *Quantum Machine Learning*

Quantum machine learning (QML) is a revolutionary technology that leverages the power of quantum computing to advance machine learning. One of the main advantages of QML is its potential to increase processing dimensions. QML has a wide range of potential applications, including the discovery and development of quantum algorithms, materials, devices, and etc. (Acampora, 2019). Recent research in QML has focused on two main approaches: search quantum methods and use of probabilistic quantum theory. Search quantum methods aim to replace classical ML algorithms with suitable quantum methods, while the other method is to use probabilistic quantum theory to explain stochastic processes (Schuld et al., 2014). Similar to quantum optimization algorithms, QML algorithms also utilizes dynamical feedback from classical evaluations that rely on the evaluation of quantum models (Fig. 2).

Several studies have demonstrated the quantum advantage in various ML applications. These studies have shown that QML can achieve quadratic gains in performance and exponential increases in performance in a short amount of time (Benedetti et al., 2018; Dunjko et al., 2016). Additionally, QML has been applied to recommendation systems (Kerenidis & Prakash, 2017), reinforcement learning (Lamata, 2017), and federated learning (Lim et al., 2021), with promising results. While QML holds great promise, there are also challenges that need to be addressed, such as collision prevention, warming issues, and networking security (Shakhatreh et al., 2019). With continued research and development, QML is poised to have a major impact on a wide range of industries and applications.

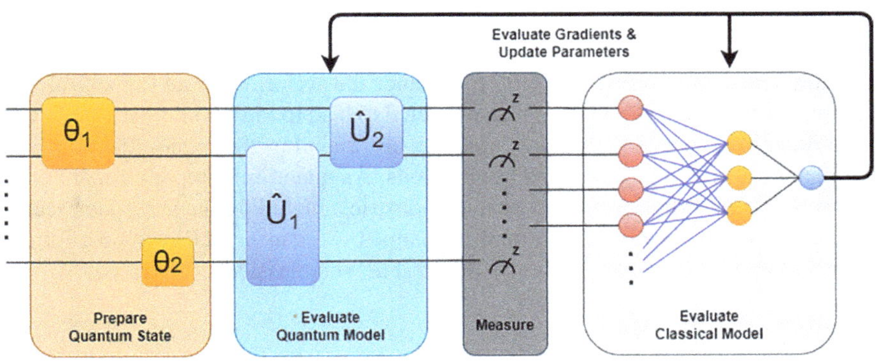

Fig. 2 Quantum machine learning

2.4 Quantum Simulations

Quantum simulations are a powerful technique used in quantum computing to simulate the behavior of complex quantum systems. In fact, it was one of the driving factors for the development of quantum devices in the early decades. Quantum simulations are primarily used for the simulation of quantum particles or atomic level structures. Two of the popular methods of quantum simulations are Quantum Phase Estimation (QPE) and Variational Quantum Eigensolver (VQE).

QPE is a widely used subroutine in quantum algorithms that assists in computing the ground state energy of a quantum computing system. It enables the measurement of almost any Hermitian operator, which is a fundamental building block employed in numerous quantum algorithms. However, due to the current state-of-the-art methods of implementation, QPE presents significant challenges in its construction, such as high quantum circuit depth (Wong, 2022a).

VQE, on the other hand, leverages both conventional and quantum computers. It facilitates the identification of values that yield the optimal solution for a given optimization problem. VQE represents a hybrid approach that synergizes the strengths of both quantum and classical realms. The central objective of this process is to ascertain the upper bound of the Hamiltonian's smallest eigenvalue.

One example of a quantum simulation algorithm is the simulation of a hydrogen molecule. This algorithm can be used to calculate the energy levels and wavefunctions of the hydrogen molecule, which are important for understanding the chemical bonding in this molecule. Another example of a quantum simulation algorithm is the simulation of a spin system. This algorithm can be used to calculate the magnetic properties of a material, which are important for understanding the material's behavior in magnetic fields.

Quantum simulations are a powerful tool for studying complex quantum systems. They have the potential to revolutionize our understanding of the physical world and lead to the development of new technologies, such as more efficient drug design and more accurate materials simulations.

As we navigate the NISQ era, we can anticipate further advancements and breakthroughs. Researchers are actively exploring novel architectures, error-correction schemes, and quantum software tools to enhance the performance and capabilities of quantum computers. These efforts are expected to lead to the development of larger-scale quantum devices with lower noise levels, enabling the exploration of more complex quantum algorithms and applications. As quantum computing matures, it has the potential to revolutionize various industries, including the aerospace sector. By harnessing the power of quantum mechanics, we can unlock new possibilities and solve problems that are currently intractable with classical computers.

3 Quantum Optimization and Data Analytics

As discussed in Sect. 2, increased autonomy necessitates advanced navigation and control capabilities. In terms of navigation, the requirement is more on accurate sensing, autonomous decision making, and efficient trajectory planning. One of the key solutions for accurate sensing is to use quantum sensors. Quantum sensing is an interesting application of quantum technologies, where quantum sensors such as quantum gravimeters have demonstrated their capability to perform accurate measurements far surpassing the already advanced inertial navigation systems currently in use. Not only quantum sensors are practically demonstrated, they are also making a transition from scientific research to real-world applications (Bongs et al., 2019). Further research is being done for feasible integration with autonomous vehicles including UAVs (Cooper et al., 2021; Earl et al., 2022). However, quantum sensing and its applications are out of the scope of this chapter.

Quantum computing itself has its own capabilities to potentially impact critical processes in navigation systems. For example, by optimizing satellite operations scheduling, planning UAV routes, and strategically allocating resources within satellite networks, quantum-based tools can directly increase overall efficiency while decreasing operating costs. Quantum algorithms further offer autonomous systems the capacity to dynamically optimize decision-making processes and correct their own pathways in real-time, minimizing dependence on ground-based commands, and empowering these systems to adapt to unexpected, dangerous, or changing circumstances (Harwood et al., 2021). Beyond individual vehicles, quantum solutions are relevant to systems design for satellite constellations or swarms of UAVs (Barbeau, 2019). Large networks like these present difficult challenges in optimizing coverage, managing signal latency, and fleet management. These developments will become crucial for mission longevity and successful long-range ventures.

3.1 Optimized Scheduling

Quantum algorithms have the potential to revolutionize satellite and UAV scheduling by considering various complex parameters simultaneously. Often, scheduling tasks is related to the Job Shop Scheduling problem (Cozzolino et al., 2021) and Portfolio optimization problems (Li et al., 2021), which are well-known NP-hard optimization problems. Unlike traditional scheduling methods, which often rely on heuristics or simplified models, quantum algorithms can leverage their inherent parallelism and superposition to find optimal solutions in a much shorter time (Desimone, 2019).

When it comes to autonomous vehicles, scheduling tasks efficiently is a major challenge. This is especially true for large fleets of drones, satellite constellations, and other autonomous systems. One organization that has been exploring the use of quantum computing to address this challenge is NASA (Guillaume et al., 2022).

NASA's Deep Space Network (DSN) consists of three main antennas, and it receives dozens of requests each week for their use. Scheduling these requests manually is a complex and time-consuming process, taking up to four months. According to NASA, "Solving the DSN scheduling problem consists in placing as many tracks as possible within a finite time span (one week) while utilizing a limited amount of resources (antennas and equipment) and satisfying other constraints." To improve the efficiency of this process, NASA has explored using a quantum annealing system to solve the DSN scheduling problem. Quantum annealing is a type of quantum computing that uses quantum effects to solve optimization problems, and the research has used quadratic unconstrained binary optimization (QUBO). In the case of the DSN scheduling problem, quantum annealing has shown to produce more efficient schedules with more request satisfactions, more resource utilization, and less idle time. This research demonstrates the potential of quantum computing to solve complex scheduling problems in the aerospace industry. As quantum computers continue to develop, they are likely to play an increasingly important role in solving a wide range of challenging problems (Guillaume et al., 2022).

Quantum algorithms can optimize the scheduling of tasks and find the most efficient routes, and ensure that all tasks are completed within the desired time frame. This can lead to major improvements in mission efficiency, operational cost reductions, and resource utilization. For example, consider fuel consumption. By optimizing the scheduling of operations, quantum algorithms can help reduce the amount of fuel needed to complete a mission. This can save organizations a lot of money in the long run.

Another benefit of using quantum algorithms is that they can help improve the utilization of resources. For example, they can help organizations find the most efficient way to allocate ground stations, communication channels, and antennas to different tasks. This can lead to increased productivity and better overall performance. This has been proven in reality with the case study of NASA's deep space network.

Overall, quantum algorithms have the potential to significantly enhance the scheduling of satellite and UAV operations, leading to increased mission efficiency, reduced operational costs, improved resource utilization, and real-time optimization capabilities.

3.2 Efficient Route Planning

Route planning is also an NP-hard problem, often with significant relationship to Traveling Salesman Problem (Matai et al., 2010). It has a range of application in UAV operations. Delivery drones must follow optimized delivery routes to ensure that they deliver the packages as quickly and efficiently as possible (Brar et al., 2015). When planning paths for military UAVs, factors such as enemy air defenses and terrain should be considered in order to improve the safety and effectiveness of military missions. While the class of problems in route planning are often hard to solve with classical computers, quantum algorithms have the potential to revolutionize route planning for unmanned aerial vehicles (UAVs). By leveraging the power of quantum computing, UAVs can generate optimal routes that consider a variety of factors.

This capability offers several advantages over traditional route planning methods. First, quantum algorithms can consider a much larger number of variables than classical algorithms, resulting in more accurate and efficient routes. Second, quantum algorithms can find optimal routes with clear runtime advantages compared to the classical algorithms (Leib et al., 2023). Third, quantum algorithms can be used to plan routes in real time, allowing UAVs to respond quickly to changing conditions. The benefits of quantum route planning for UAVs are significant. Optimized routes can minimize flight time, fuel consumption, and exposure to potential hazards. This can lead to increased mission efficiency, cost savings, and increased safety (Harwood et al., 2021).

3.3 Network Optimization

Quantum computing can be utilized to design optimal network architectures for satellite constellations or UAV swarms. These networks are becoming increasingly important for various applications, including communication, Earth observation, and defense. Quantum algorithms can consider multiple factors such as coverage, communication latency, and security requirements to determine the optimal placement and configuration of nodes in the network.

For example, Kim J. et al. suggests modelling a satellite constellation scheduling problem as a max-weight independent set problem, which could then be solved using quantum optimization-based algorithms (Kim et al., 2021). The problem in interest is to reduce having server overlapped monitoring areas among satellites. The problem of overlapping area scheduling was mapped to a max-weight independent set problem and then suggested of solving using QAOA.

Optimized network architectures achieved through quantum computing can significantly enhance overall network performance. By optimizing the network topology, resource allocation, and routing protocols, quantum computing can improve communication reliability, reduce latency, and increase network capacity. This can lead to improved user experience, enhanced efficiency, and cost savings. This has been highlighted in a proposal to use genetic algorithms for networks of large-scale UAV swarms (Mousavi et al., 2019). As networks continue to grow in size and complexity, quantum algorithms can efficiently handle the increased number of variables and constraints involved in network optimization. Quantum-optimized networks can also be easily adapted to changing network conditions, such as traffic patterns, node failures, and security threats.

Overall, network optimization using quantum computing holds great promise for transforming the way networks are designed and managed. By leveraging the power of quantum computing, organizations can achieve unprecedented levels of network performance, reliability, scalability, flexibility, and security.

3.4 Data Processing and Analytics

As the tasks aboard spacecraft and drones increase in complexity, the importance of advanced computational power escalates. Classical computers might struggle with the demand for real-time data processing, sophisticated pattern recognition, and handling the overwhelming torrent of data gathered by modern sensors due to their inherited computational limitations. However, the emerging fields like QML suggests that quantum algorithms excel at pattern finding in massive datasets (Schuld & Petruccione, 2021). Due to the speed-ups that quantum algorithms facilitate in machine learning, quantum algorithms could be used for effective image analysis, target recognition, and a real-time, high-resolution understanding of a constantly changing environment (Shaikh & Ali, 2016).

While many of the previously discussed problems were given quantum solutions in different industries, QML has been proven with solutions in drone applications. QML has been used to address challenges in drone applications, such as object-detection, visual identification, and data analysis. ML algorithms have been implemented to optimize drone placements and improve wireless systems (Li et al., 2020).

The advanced data analysis capabilities of QML empower organizations to proactively identify and address potential issues in their satellite and UAV systems. By leveraging QML techniques, organizations can minimize operational disruptions, improve system reliability, and gain a competitive edge in the aerospace industry (IBM Institute for Business Value, 2023).

As QML continues to evolve, we can expect even more advanced and transformative applications in the field of satellite and UAV data analysis. These techniques hold the promise of revolutionizing industries such as remote sensing, environmental monitoring, and precision agriculture, unlocking new possibilities for data-driven decision-making and innovation.

4 Quantum Communication

Quantum computing has both positive and negative effects on cybersecurity. On the positive side, quantum computing enhances security with quantum communication and QKD, which can provide unconditional security against eavesdropping. Quantum computing can also improve security by enabling advanced machine learning capabilities, which can help design more secure systems, prevent cyberattacks, and recover from cyberattacks. On the negative side, quantum computing can also break security by threatening existing encryption schemes, such as RSA and ECC, which are widely used to protect data and communications on the internet (Suo et al., 2020).

Several solutions and countermeasures such as quantum-resistant cryptography, quantum-proof protocols, quantum detection, and QKD have been proposed and developed to mitigate the negative effects of quantum computing on cybersecurity.

4.1 Quantum-Resistant Encryption

At the heart of the quantum threat lies Shor's algorithm, a quantum computing breakthrough that has the potential to undermine widely used public-key cryptographic systems such as RSA and ECC (Suo et al., 2020). These widely used protocols form the backbone of secure communications and digital signatures. Currently, breaking these cryptographic schemes would take conventional computers an impossibly long time to carry out the factoring. Yet, Shor's algorithm, operating on a sufficiently powerful quantum computer, could efficiently crack them.

Post-quantum cryptography (PQC) is the solution to this emerging problem. PQC is an active field of research focused on encryption algorithms designed with inherent mathematical resistance to attacks even from large-scale quantum computers. Examples of quantum-resistant cryptography include lattice-based cryptography, code-based cryptography, hash-based cryptography, etc.

In 2016, the US National Institute of Standards and Technology (NIST), initiated a PQC Standardization process where researchers, universities and companies were publicly invited to develop cryptographic schemes that are secure from both classical and quantum threats (NIST, 2016). As per the initiative, four protocols were announced as reliable post-quantum encryption standards. These includes CRYSTALS-Kyber public key encryption, CRYSTALS-Dilthium digital signature algorithm, Falcon digital signature algorithm, and SPHINCS+ digital signature algorithm (IBM, 2023). These approaches to encryption differ in their strengths, weaknesses, and levels of maturity, prompting ongoing evaluation and standardization efforts.

4.2 Protecting Sensitive Data on Satellites and UAVs

Satellites and UAVs play a major role in both communication and reconnaissance sectors. They collect, generate, and store vast troves of sensitive data, ranging from high-resolution imagery to mission-critical sensor readings and telemetry. As quantum computing capabilities develop, it is essential to protect this data from unauthorized access, tampering, and interception.

To shield critical systems from quantum attacks, a proactive strategy is pivotal. Adopting and integrating post-quantum encryption methods across data transmission, storage protocols, and digital signatures within satellites and UAVs secures these devices through every stage of data handling.

While advanced encryption is important, the battle against quantum-enabled cyber threats extends beyond algorithm replacements. Classical information has no restrictions against being copied or stored. However, quantum information doesn't allow for copying of information which stems from the no-cloning theorem of quantum information (Wong, 2022b). Therefore, quantum communication provides unconditional security against eavesdropping and hacking attacks, making it ideal for transmitting sensitive data. This technology enables secure and reliable communication in space, ensuring the confidentiality and integrity of transmitted information.

However, implementing these new PQC algorithms and quantum communication channels involves substantial architectural reconfiguration and careful management. The transition to quantum-resistant encryption protocols cannot happen instantly—thorough testing and gradual integration will be needed for a smooth and secure migration. Transferring to quantum communication is even more challenging due to the hardware and infrastructure requirements.

4.3 *Quantum Cryptography*

Quantum cryptography, widely known as Quantum Key Distribution (QKD), is a method of random key generation and distribution. QKD marks a turning point in guaranteeing information security. Instead of conventional encryption that might be vulnerable to sophisticated attacks (especially concerning for future quantum computers), QKD exploits the fundamental laws of quantum physics such as no-cloning theorem, uncertainty principle, and entanglement to ensure secure communication links. Using QKD protocols, any attempt to eavesdrop on exchanged keys introduces detectable disturbances, safeguarding data privacy and rendering communication links resistant even to powerful cyber threats (Wolf, 2021).

In aerospace, satellites, ground stations, and UAVs could reap the benefits of establishing truly secure channels of communication. Ground-to-satellite or satellite-to-satellite QKD networks could pave the way for ultra-secure military communications, protection of sensitive government data, and secure command links for satellite systems. Additionally, QKD could underpin secure communications within and between swarms of UAVs. As this is an important technology with many real-world demonstrations, the next section is dedicated to understanding what QKD is.

4.4 *Quantum Key Distribution*

The foundation of QKD relies on two fundamental quantum principles. One is the uncertainty principle. In the previous sections, we learned that quantum particles can exist in multiple states simultaneously, which is known as superposition. The quantum uncertainty principle states that the exact state of the particle cannot be verified without performing a measurement. The other principle is the no-cloning theorem, which states that it's impossible to create an exact copy of an unknown quantum state. Any attempt to measure a quantum system inherently disturbs its original state, leaving a detectable trace.

However, measurement of a superposition state collapses its quantum state to a classical state. The result of this phenomena is if a person (Alice) sends some information by encoding them into a superposition state of a qubit to another person (Bob), any attempt to eavesdrop requires performing a measurement, thereby collapsing the initial state and introducing an error. This fundamental aspect in nature allows us to create communication channels that are resistant to eavesdropping attacks.

Before diving further into QKD, it's important to also look at the nature of existing classical cryptography schemes and how QKD compares to such encryption schemes.

In classical cryptography, there are two main categories of secret key distribution: Symmetric Key Distribution (also known as Secret Key Cryptography) and Asymmetric Key Distribution (Public Key Cryptography).

Both of these schemes require a key to encrypt data and decrypt previously encrypted data. The fundamental difference is, in symmetric key cryptography the two keys are equivalent while the asymmetric key cryptography requires two different keys. Symmetric key cryptography is similar to two people meeting at a room and sharing a pass-key for future use, which they will use for their communication afterwards. Both the sender of the message and the receiver uses the same key for encryption and decryption. However, sharing the secret key over an exposed channel is highly vulnerable.

In asymmetric key cryptography, sender and receiver uses different keys and combine them with a mathematical function to create their public keys. When sending data over a communication channel sender encrypts the data using their public key and the receiver decrypts it using their private key. The mathematical functions are carefully designed to make this possible while also making the public keys (as the name suggests) publicly available, without making the communication vulnerable.

These mathematical functions in asymmetric key systems such as RSA rely on prime factorization, which is given a large enough number is a computationally impractical problem for classical computers, but a solvable problem for a large quantum system. This is the widely discussed quantum threat that exist for encryption.

However, symmetric key cryptography systems such as one-time pads are theoretically perfectly safe against any kind of cyber-attacks if the keys could be shared without exposal. This is where QKD sheds a new light (Fig. 3). Quantum systems, based on the previously discussed facts, are capable of facilitating communication channels that are safe from eavesdropping. Therefore, quantum communication channels could be used to make these quantum one-time pads a reality, which has become known as QKD.

4.5 QKD Experiments

QKD technology has experienced significant development in recent years. Unlike many other quantum technologies, QKD has proven its applicability to existing commercial systems and has a track record of successful experiments showcasing its capabilities.

The first demonstration of a QKD network was achieved in 2003 in Cambridge, Massachusetts, under the sponsorship of DARPA (Elliott, 2002). This 29-km network connected Harvard University, BBN Technologies, and Boston University, utilizing

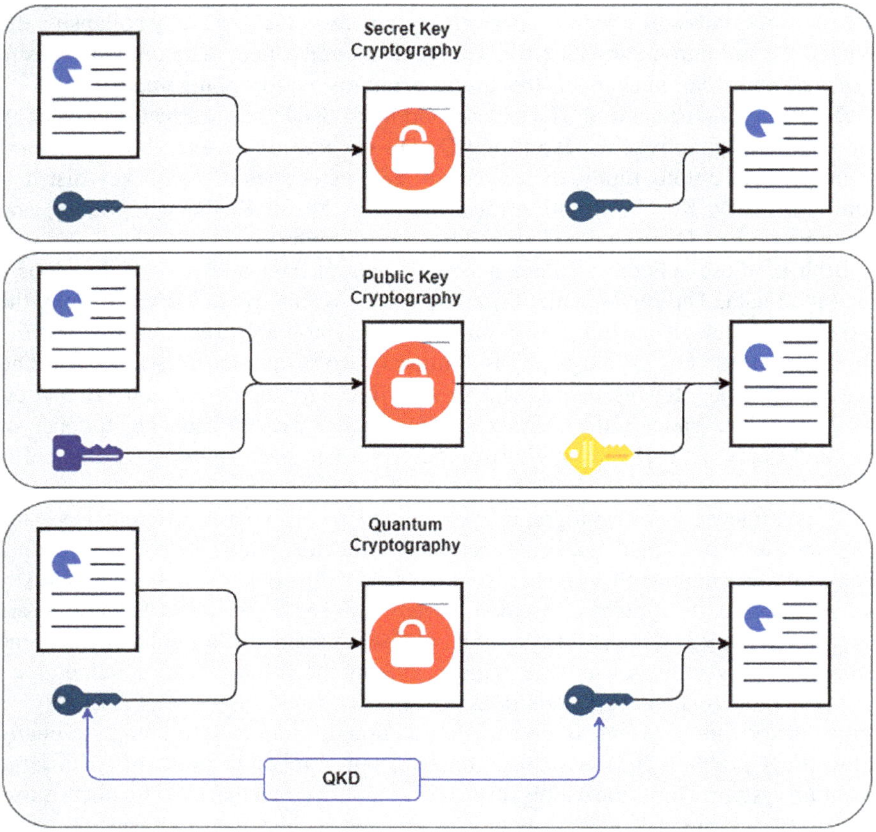

Fig. 3 Classical vs. quantum cryptography

standard telecommunication fiber strands (Elliott, 2005). In 2007, the first commercial QKD system was installed between Vienna and Innsbruck, Austria. This system employed fiber optic cables to distribute quantum keys over a distance of 144 km (Peev et al., 2009). In 2021, a significant milestone was reached when a team of researchers achieved a record-breaking transmission distance of 500 km using a fiber optic cable. This achievement was an important demonstration for the potential of QKD for long-distance secure communication (Chen et al., 2021).

QKD has also been successfully tested in space-based scenarios. Several countries have launched satellites or conducted experiments on the International Space Station (ISS) to demonstrate QKD protocols. These experiments have shown the ability to generate unbreakable encryption keys over satellite-ground and even satellite-to-satellite links.

In 2016, China's launch of the Micius satellite marked a landmark achievement in space-based QKD. This satellite demonstrated satellite-to-ground QKD over a distance of 1200 km, establishing secure video links between Beijing and

Vienna. This successful experiment not only proved the feasibility of long-distance quantum communication in space but also highlighted its potential impact on secure international communication (Liao et al., 2017).

The Canadian Space Agency (CSA) is also actively pursuing space-based QKD through their Quantum Encryption and Science Satellite (QEYSSat) mission. This project aims to demonstrate secure links and test the building blocks of future distributed quantum networks spanning space and ground nodes (Canadian Space Agency [CSA], 2023).

Additionally, QKD is not solely confined to large satellites. Research is underway to explore QKD via smaller satellites (Miller et al., 2023) and drones (Conrad, 2021), offering potential solutions for secure communication in various scenarios.

5 Challenges

The field of quantum computing has the potential to revolutionize various industries, including the aerospace sector. However, there are several challenges that need to be overcome before quantum computing can be used in space and aerial vehicles. The first major challenge in applying quantum computing for space and aerial vehicles is the size of current quantum systems. Today's quantum computers are analogous to the early classical computers that required large rooms, consumed significant power, and had limited performance.

The size of current quantum systems poses several challenges for their integration into space and aerial vehicles. Spacecraft and aircraft have strict size, weight, and power constraints. Quantum systems that are too large or consume too much power may not be feasible for these applications. Additionally, the fragile nature of quantum systems requires specialized environmental conditions, such as low temperatures and vibration isolation, which may be difficult to maintain in the harsh environments of space or during flight.

The harsh environment of space radiation can disrupt quantum hardware. Today's quantum computers are considered highly sensitive devices and must undergo radiation hardening to ensure their resilience against the threats posed by radiation on Earth (Bushnell, 2021). Radiation hardening for space could become an enormous challenge. Similar radiation environments apply for high-altitude surveillance UAVs. Therefore, on-board quantum systems for UAVs and satellites should be either quantum devices that are tolerant to radiation or devices that are perfectly radiation hardened. However, current technological capabilities in radiation hardening requires such highly sensitive quantum devices to be surrounded by equipment of considerably large physical dimensions.

Similar challenges are faced due to the thermal environments. Quantum computers generate heat, which can affect their stability and performance. Effective thermal management systems are required to maintain the optimal operating temperature for quantum computers. Today's systems that facilitate this thermal management too are of considerably large physical dimensions.

Another major challenge faced by quantum computing is noisy qubits. Qubits are inherently vulnerable to errors caused by noise and decoherence. This can lead to incorrect results in computations, which is why it has become difficult to obtain quantum supremacy in NISQ systems. Complex error correction schemes are essential for scaling quantum computers, as they allow for the detection and correction of errors. The development of effective error correction methods is a major research focus. It will determine the practical feasibility of large-scale quantum computers. These advancements will have profound implications for space-based applications, enabling more powerful and reliable quantum computing (Roffe, 2019).

Current quantum algorithms are capable of handling small-scale problems effectively. However, for quantum computing to have a transformative impact in space, it is necessary to scale up existing algorithms. This represents a significant challenge, as it requires the development of new algorithms that can efficiently utilize the increased computational power of larger quantum computers. Overcoming this hurdle is crucial for realizing the full potential of quantum computing in space (Pascuzzi et al., 2022).

Addressing these challenges will pave the way for the integration of quantum computing into space missions, enabling a new era of scientific discovery and technological innovation.

6 Governance, Policy, and Standardization

Quantum technologies promise revolutionary advancements for space and aerial vehicles. However, their integration into these domains necessitates comprehensive governance to ensure their responsible and ethical use. This requires addressing multifaceted legal, ethical, and regulatory challenges.

With the rapid immergence of AI, we have witnessed the importance of parallel advancement of governance and policies with the respective technologies. Some claim that it is too late to govern AI because proper governance of AI technologies within the national governing bodies and laws are too slow to adapt to the rapidly changing landscape of AI (Sundaram, 2024). Therefore, it is important to highlight the advancements in the quantum computing governance while talking about the technical aspects of quantum computing.

There has been a considerable effort from researchers to highlight the importance of governing quantum technologies and the legal and ethical challenges related to them. One such comprehensive framework is provided by Mauritz by proposing ten guidelines for the development and applications of quantum technologies (Kop, 2021). Another comprehensive analysis is provided in the insight report by World Economic Forum (Coates et al., 2022). It defines governance in terms of core values such as common good, accountability, inclusiveness, equitability, non-maleficence, accessibility, and transparency. Further, it identifies nine driving themes of quantum developments: Transformative capabilities, access to hardware infrastructure, open innovation, creating awareness, workforce development and capacity-building, cybersecurity, privacy, standardization, and sustainability. Furthermore,

Mitchell Institute for Aerospace Studies details a three-part policy paper on responsible use of quantum advantage (Penney, 2024). The US government and many other governments in the European union have introduced their development plans to their state laws (Public Law No: 115–368, 2018).

The management of quantum technologies in aerospace involves crafting regulatory frameworks that consider various interconnected domains. Data privacy, national security, export controls, intellectual property rights, and societal effects all play a crucial role.

Quantum computing's rapid pace poses challenges to current data privacy and security practices. As quantum computers threaten existing encryption methods, adopting quantum-resistant encryption standards, secure communication protocols, and robust key management systems is essential to maintain confidentiality. Given the extensive datasets generated by quantum systems on aerial and space platforms, reviewing and establishing new data handling, storage, and transmission standards is paramount.

International collaboration is imperative for effective governance of quantum technologies in space. The multinational nature of space flight and the growing accessibility of aerial drone technology necessitate shared regulatory standards. Collaborative efforts among space agencies, private industries, and research institutions across borders are crucial to prevent the dangers of uncontrolled technological advancements. This includes developing globally accepted technical standards for secure communication, reliable encryption, and benchmarks for quantum sensor accuracy and calibration.

The integration of quantum technologies into air and space vehicles presents not only technical challenges. Ethical frameworks must be actively developed in parallel with technological advancements to ensure responsible usage, mitigate potential risks, promote fairness, and maintain accessibility. The extended capabilities of quantum-enhanced surveillance and monitoring requires transparent societal discussions that finds a balance between enhanced security and the protection of individual privacy.

Safety and risk assessments are essential before widespread implementation. A comprehensive evaluation of the effects on existing satellite and aerial systems, the safety of humans, and the sustainability of the environment is crucial. Early identification of risks and the development of mitigation strategies are vital components of responsible deployment.

Governance of quantum technologies in this domain requires a multi-faceted approach. Lawmakers, engineers, pilots, and the general public should receive continuous education on the capabilities, risks, and ethical implications of these technologies. Informed discussions and decision-making processes involving various stakeholders are essential alongside technical innovation to foster an environment where ethical considerations remain at the forefront.

Therefore, a dynamic approach is necessary to keep pace with the rapid evolution of quantum technologies. Regulation must be flexible and forward-looking, providing timely safeguards while allowing the necessary space for further innovation. The only way to safely utilize the benefits that quantum technology offers is through a collaborative, internationally coordinated effort.

7 Conclusion

Quantum technologies hold immense potential for space and aerial vehicle applications, promising improved efficiency, security, and forecasting capabilities. However, effective governance is essential to ensure responsible and beneficial utilization. By addressing data privacy, ethical considerations, safety, and international collaboration, governments can foster innovation and sustainable growth in the space and aerial vehicle industries during the quantum era.

In the previous sections we looked at the promises of quantum computing to revolutionize how we design, operate, and utilize space and aerial vehicles. It was clearly evident that quantum technologies offer the potential to revolutionize space and aerial vehicles through enhanced navigation, secure communication, and accelerated data processing.

Quantum computing can address challenges such as large-scale system coordination, trajectory optimization, processing large data volumes, and simulating quantum systems. Quantum optimization algorithms, like QAOA, exploit quantum parallelism to find optimal solutions efficiently. Quantum computers can also accelerate machine learning tasks and contribute to breakthroughs in industries like material development. Enhanced navigation powered by quantum sensors could offer unmatched precision, crucial in space or dense urban environments. Quantum communication, particularly QKD, promises unbreakable encryption for secure data transmission. However, quantum computing is still in its early stages, with challenges like noisy quantum devices and the need for specialized algorithms.

The implications of quantum-enabled aerospace technology are far-reaching. Satellites may achieve new levels of autonomy, coordinating efficiently in constellations and adapting to changing conditions. UAVs could navigate complex airspaces safely and reliably, expanding their roles in transportation, delivery, and surveillance. Quantum computing could also accelerate the design process, allowing for the simulation and optimization of new aerial and space vehicles with unmatched speed and detail.

The quantum era of space and aerial vehicles necessitates a multidisciplinary approach. Quantum physicists, engineers, computer scientists, and policy experts must collaborate seamlessly. Quantum physicists lay the foundation by deepening our understanding, engineers translate principles into robust systems, computer scientists develop algorithms, and policy experts ensure responsible deployment. Funding, algorithm refinement, and collaboration are cornerstones of progress. Responsible utilization is key and involves addressing data privacy, ethics, safety, and international collaboration for standardized approaches. Success will enable us to harness the incredible potential of quantum-enabled space and aerial vehicles while upholding responsible and equitable utilization.

The focus of the chapter has been to highlight the transformative potential of quantum computing while emphasizing the need for continued collaboration across disciplines. The future of space and aerial vehicles depends on exploring disruptive technologies. Informed and responsible efforts can build a future where quantum technologies aid in navigating the cosmos and safeguarding skies.

References

Abdelkader, M., Güler, S., Jaleel, H., & Shamma, J. S. (2021). Aerial Swarms: Recent Applications and Challenges. *Current Robotic Reports, 2*, 309–320.

Acampora, G. (2019). Quantum Machine Intelligence. *Quantum Machine Intelligence, 1*, 1–3.

Applin, S. A. (2016). Deliveries by Drone: Obstacles and Sociability. In B. Custers (Ed.), *The Future of Drone Use: Information Technology and Law Series* (Vol. 27, pp. 71–91). T.M.C. Asser Press.

Arute, F., Arya, K., Babbush, R., et al. (2019). Quantum Supremacy Using a Programmable Superconducting Processor. *Nature, 574*(7779), 505–510.

Barbeau, M. (2019). Recognizing Drone Swarm Activities: Classical versus Quantum Machine Learning. *Digitale Welt, 3*, 45–50.

Benedetti, M., Realpe-Gómez, J., & Perdomo-Ortiz, A. (2018). Quantum-Assisted Helmholtz Machines: A Quantum–Classical Deep Learning Framework for Industrial Datasets in Near-Term Devices. *Quantum Science and Technology,* 3(3).

Bongs, K., Holynski, M., Vovrosh, J., Bouyer, P., et al. (2019). Taking Atom Interferometric Quantum Sensors from Laboratory to Real-World Applications. *Nature Reviews Physics, 1*, 731–739.

Bongs, K., Bennett, S., & Lohmann, A. (2023). Quantum Sensors Will Start a Revolution—If We Deploy Them Right. *Nature, 617*, 672–675.

Brar, S., Rabbat, R., Raithatha, V., Runcie, G., & Yu, A. (2015). *Drones for Deliveries.* Sutardja Center for Entrepreneurship & Technology.

Bushnell, D. M. (2021). *Futures of Deep Space Exploration, Commercialization, and Colonization: The Frontiers of the Responsibly Imaginable.* NASA Langley Research Center.

Canadian Space Agency. (2023, July 118). *Quantum Encryption and Science Satellite (QEYSSat).* Canadian Space Agency (CSA). [Online]. https://www.asc-csa.gc.ca/eng/satellites/qeyssat.asp. Accessed 25 Feb 2024.

Chen, J., Zhang, C., Liu, Y., et al. (2021). Twin-Field Quantum Key Distribution Over a 511 Km Optical Fiber Linking Two Distant Metropolitan Areas. *Nature Photon, 15*, 570–575.

Coates, R., Dayal, R., Gagliardoni, T., Greplová, E., Humble T., et al. (2022). *Quantum Computing Governance Principles.* World Economic Forum.

Conrad, A., Isaac, S., Cochran, R., Sanchez-Rosales, D., Wilens, B., Gutha, A., Rezaei, T., Gauthier, D. J., & Kwiat, P. (2021). Drone-Based Quantum Key Distribution (QKD). In *Proceedings Volume 11678, Free-Space Laser Communications XXXIII*. (Online).

Cooper, N., Coles, L., Everton, S., Campion, I. M. R., Madkhaly, S., et al. (2021). Additively Manufactured Ultra-High Vacuum Chamber for Portable Quantum Technologies. *Additive Manufacturing, 40*, 101898.

Cozzolino, C., Christiansen, B., Vallejos, E., Sorce, M., & J. Visk. (2021, December 15). *Job Shop Scheduling.* Cornell University Computational Optimization Open Textbook. [Online]. https://optimization.cbe.cornell.edu/index.php?title=Job_shop_scheduling. Accessed 20 Feb 2024.

Desimone, R. (2019, March). *Quantum Computing Algorithms for Optimised Planning & Scheduling (QCAPS).* [Online]. https://www.dwavesys.com/media/ru5atj5p/11_qcaps-qubits-europe-march-2019.pdf. Accessed 20 Feb 2024.

Dunjko, V., Taylor, J. M., & Briegel, H. J. (2016). Quantum-Enhanced Machine Learning. *Physical Review Letters,* 117(13).

Earl, L., Vovrosh, J., Wright, M., Roberts, D., Winch, J., Perea-Ortiz, M., et al. (2022). Demonstration of a Compact Magneto-Optical Trap on an Unstaffed Aerial Vehicle. *Atoms, 10*, 32.

Elliott, C. (2002). Building the Quantum Network. *New Journal of Physics, 1*, 46.

Elliott, C. (2005). *The DARPA Quantum Network*. arXiv.

Elmokadem, T., & Savkin, A. V. (2021). Towards Fully Autonomous UAVs: A Survey. *Sensors (Basel), 21*(18), 6223.

Engberts, B., & Gillissen, E. (2016). Policing from Above: Drone Use by the Police. In B. Custers (Ed.), *The Future of Drone Use: Information Technology and Law Series* (Vol. 27, pp. 93–113). T.M.C. Asser Press.

Estrada, M. A. R., & Ndoma, A. (2019). The Uses of Unmanned Aerial Vehicles—UAV's-(or Drones) in Social Logistic: Natural Disasters Response and Humanitarian Relief Aid. *Procedia Computer Science, 149*, 375–383.

Farhi, E., Goldstone, J., & Gutmann, S. (2014). *A Quantum Approximate Optimization Algorithm*. arXiv preprint. arXiv:1411.4028

Gisin, N., & Thew, R. (2007). Quantum Communication. *Nature Photonics, 1*(3), 165–171.

Grover, L. K. (1996). A Fast Quantum Mechanical Algorithm for Database Search. In *STOC '96: Proceedings of the Twenty-Eighth Annual ACM Symposium on Theory of Computing.*

Guillaume, A., Goh, E. Y., Johnston, M. D., Wilson, B. D., Ramanan, A., Tibble, F., & Lackey, B. (2022). Deep Space Network Scheduling Using Quantum Annealing. *IEEE Transactions on Quantum Engineering, 3*, 1–13.

Harwood, S., Gambella, C., Trenev, D., Simonetto, A., Bernal, D., & Greenberg, D. (2021). Formulating and Solving Routing Problems on Quantum Computers. *IEEE Transactions on Quantum Engineering, 2*, 1–17.

IBM. (2023, February 17). *What Is Quantum-Safe Cryptography?* IBM. [Online]. https://www.ibm.com/topics/quantum-safe-cryptography. Accessed 23 Feb 2024.

IBM Institute for Business Value. (2023). *The Quantum Decade*. IBM Corporation.

Johnson, T. H., Clark, S. R., & Jaksch, D. (2014). What Is a Quantum Simulator? *EPJ Quantum Technology, 1*, 10.

Kerenidis, I., & Prakash, A. (2017). Quantum Recommendation Systems. In *8th Innovations in Theoretical Computer Science Conference (ITCS 2017).*

Kim, J., Kwak, Y., Jung, S., & Kim, J.-H. (2021). *Quantum Scheduling for Millimeter-Wave Observation Satellite Constellation*. arXiv:2108.00626 [eess.SY].

Kop, M. (2021). Establishing a Legal-Ethical Framework for Quantum Technology. *Yale Law School, Yale Journal of Law & Technology (YJoLT).*

Lamata, L. (2017). Basic Protocols in Quantum Reinforcement Learning with Superconducting Circuits. *Science Reports, 1609.*

Lee, C. (2016). *The Culture of US Air Force Innovation*. King's College.

Leib, D., Seidel, T., Jäger, S., et al. (2023). An Optimization Case Study for Solving a Transport Robot Scheduling Problem on Quantum-Hybrid and Quantum-Inspired Hardware. *Science Reports, 13*, 18743.

Li, Y., Aghvami, A. H., & Dong, D. (2020). *Intelligent Trajectory Planning in UAV-Mounted*. arXiv, vol. [eess.SP].

Li, F., Pan, K., Sun, Q., Li, H., & Luo, E. (2021, December 17). *Portfolio Optimization*. Cornell University Computational Optimization Open Textbook. [Online]. https://optimization.cbe.cornell.edu/index.php?title=Portfolio_optimization. Accessed 20 Feb 2024.

Liao, S., Cai, W., Liu, W., et al. (2017). Satellite-to-Ground Quantum Key Distribution. *Nature, 549*, 43–47.

Lim, W. Y. B., Huang, J., Xiong, Z., Kang, J., Niyato, D., Hua, X.-S., Leung, C., & Miao, C. (2021). Towards Federated Learning in UAV-Enabled Internet of Vehicles: A Multi-Dimensional Contract-Matching Approach. *IEEE Transactions on Intelligent Transportation Systems, 22*, 5140–5154.

Little, B. (2019, March 20). *Automation of Planes Began 9 Years After the Wright Bros Took Flight—But It Still Leads to Baffling Disasters*. A&E Television Networks. [Online]. https://www.history.com/news/plane-automation-autopilot-flight-302-610. Accessed 16 Feb 2024.

Marzena Półka, S. P. Ł. K. (2017). The Use of UAV's for Search and Rescue Operations. *Procedia Engineering, 192*, 748–752.

Matai, R., Singh, S., & Mittal, M. L. (2010). Traveling Salesman Problem: An Overview of Applications, Formulations, and Solution Approaches. In D. Davendra (Ed.), *Traveling Salesman Problem, Theory and Applications*. IntechOpen.

McGeoch, C. C., Harris, R., Reinhardt, S. P., & Bunyk, P. (2019). *Practical Annealing-Based Quantum Computing*. D-Wave Systems Inc.

Miller, A. V., Pismeniuk, L. V., Duplinsky, A. V., Merzlinkin, V. e., Aleksandr, A., et al. (2023). Vector—Towards Quantum Key Distribution with Small Satellites. *EPJ Quantum Technology, 10*, 52.

Mohsan, S. A. H., Othman, N. Q. H., Li, Y., Alsharif, M. H., & Khan, M. A. (2023). Unmanned Aerial Vehicles (UAVs): Practical Aspects, Applications, Open Challenges, Security Issues, and Future Trends. *Intelligent Service Robots, 16*, 109–137.

Mousavi, S., Afghah, F., Ashdown, J. D., & Turck, K. (2019). Use of a Quantum Genetic Algorithm for Coalition Formation in Large-Scale UAV Networks. *Ad Hoc Networks, 87*, 26–36.

Nielsen, M. A., & Chuang, I. L. (2000). *Quantum Computation and Quantum Information*. Cambridge University Press.

NIST. (2016, December 19). *Announcing Request for Nominations for Public-Key Post-Quantum Cryptographic Algorithms*. NIST. [Online]. https://csrc.nist.gov/news/2016/public-key-post-quantum-cryptographic-algorithms. Accessed 25 Feb 2024.

Pascuzzi, V. R., Bao, N., & Li, A. (2022). *On the Importance of Scalability and Resource Estimation of Quantum Algorithms for Domain Sciences*. arXiv [quant-ph].

Peev, M., Pacher, C., Alléaume, R., Barreiro, C., Bouda, J., Boxleitne, W., Debuisschert, T., Diamanti, E., Dianati, M., & Dynes, J. F. (2009). The SECOQC Quantum Key Distribution Network in Vienna. *New Journal of Physics, 11*, 075001.

Penney, H. R. (2024). The Quantum Advantage: Why It Matters and Essential Next Steps (Part 1,2,3). *Mitchell Institute Policy Paper, 46*(47), 48.

Preskill, J. (2018). Quantum Computing in the NISQ Era and Beyond. *Quantum, 2*, 79.

Public Law No: 115–368. (2018, December 21). *H.R.6227—National Quantum Initiative Act*. US Congress.

Qiskit Optimization Development Team. (2024, February 20). *Max-Cut and Traveling Salesman Problem*. [Online]. https://qiskit-community.github.io/qiskit-optimization/tutorials/06_examples_max_cut_and_tsp.html. Accessed 23 Feb 2024.

Roffe, J. (2019). Quantum Error Correction: An Introductory Guide. *Contemporary Physics, 60*(3), 226–245.

Schuld, M., & Petruccione, F. (2021). *Machine Learning with Quantum Computers* (2nd ed.). Springer.

Schuld, M., Sinayskiy, I., & Petruccione, F. (2014). An Introduction to Quantum Machine Learning. *Contemporary Physics, 56*(2), 172–185.

Sepulveda, J. (2022, November 30). *What Is the Quantum Technology*. Airbus. [Online]. https://securecommunications.airbus.com/en/newsroom/stories/what-is-the-quantum-technology. Accessed 16 Feb 2024.

Shaikh, T. A., & Ali, R. (2016). Survey, Quantum Computing in Big Data Analytics: A Survey. In *IEEE International Conference on Computer and Information Technology (CIT)*.

Shakhatreh, H., Sawalmeh, A. H., Al-Fuqaha, A., Dou, Z., Almaita, E., Khalil, I., Othman, N. S., Khreishah, A., & Guizani, M. (2019). Unmanned Aerial Vehicles (UAVs): A Survey on Civil Applications and Key Research Challenges. *IEEE Access, 7*, 48572–48634.

Shor, P. (1997). Polynomial-Time Algorithms for Prime Factorization and Discrete Logarithms on a Quantum Computer. *SIAM Journal on Computing, 26*(5), 1484–1509.

Spremo, S. M., Crocker, A. R., & Panontin, T. L. (2019). *Small Spacecraft Overview*. NASA Ames Research Center.

Suo, J., Wang, L., Yang, S., Zheng, W., & Zhang, J. (2020). Quantum Algorithms for Typical Hard Problems: A Perspective of Cryptanalysis. *Quantum Information Processing, 19*, 1–26.

Sundaram, V. (2024, January 18). AI: Who's Governing It, and Who Should Be? *Forbes*. [Online]. https://www.forbes.com/sites/forbesbusinessdevelopmentcouncil/2024/01/18/ai-whos-governing-it-and-who-should-be/?sh=5f816765cfb6. Accessed 25 Feb 2024.

Vandersypen, L. M. K., Steffen, M., Breyta, G., Yannoni, C. S., Sherwood, M. H., & Chuang, I. L. (2001). Experimental Realization of Shor's Quantum Factoring Algorithm Using Nuclear Magnetic Resonance. *Nature, 414*, 883–887.

Wolf, R. (2021). *Quantum Key Distribution* (1st ed.). Springer.

Wong, H. (2022b). No-Cloning Theorem and Quantum Teleportation I. In *Introduction to Quantum Computing*, Springer.

Wong, H. (2022a). Quantum Phase Estimation. In *Introduction to Quantum Computing*. Springer. https://doi.org/10.1007/978-3-030-98339-0_28

Secure and Resilient IP-Satellite Communication Infrastructure

Krison Hasanaj, Osama Akram Amin Metwally Hussien, and Hamid Jahankhani

Abstract Ground-based satellite operations play a pivotal role in ensuring the seamless functioning of satellites orbiting our planet. These operations, however, are not immune to the ever-evolving landscape of cybersecurity risks and threats with the rise of quantum computing. The integrity, confidentiality, and availability of critical satellite data and services are at stake. In response to this challenge, this chapter endeavors to construct a robust cryptographic framework for mitigating the cybersecurity risks associated with the current standard cryptographic algorithms employed in IP communication for ground-based satellite operations. These algorithms, while widely used, face vulnerabilities that can be exploited by malicious actors when quantum computing become mainstream, our investigation aims to identify a promising candidate protocol that can effectively address these risks. This protocol should exhibit resilience against cyber threats from quantum computing, ensuring the secure transmission of data between ground stations and satellites. By doing so, we pave the way for a more secure and resilient IP-satellite communication infrastructure, safeguarding critical operations in an era where cybersecurity is paramount.

Keywords IP-satellite · Quantum computing · Communication · Ground based satellite stations · Cyber threats · AI · Cryptographic algorithms

1 Introduction

Current telecommunications, remote sensing, navigation, and scientific study increasingly rely considerably on ground-based satellite operations. The safety and quality of these activities are increasingly important as our dependence on satellite systems grows. However, cybersecurity risks and attacks are becoming more and more likely to affect ground-based satellite operations, which could have significant consequences for critical facilities, national security, and public safety. The

K. Hasanaj · O. A. A. M. Hussien · H. Jahankhani (✉)
Northumbria University, London, UK
e-mail: Hamid.jahankhani@northumbria.ac.uk

H. Jahankhani et al. (eds.), *Space Governance*, Space Law and Policy,
https://doi.org/10.1007/978-3-031-62228-1_5

creation of a thorough methodology that accurately assesses the cybersecurity risks and dangers connected to ground-based satellite operations has become essential.

Satellite systems face many cyber risks due to the fast development of information technology and the high interconnectivity of ground-based operations. Some people might try to attack satellite systems to take advantage of their weaknesses, stop their operations, and get their sensitive data. These cyber-attacks can use different methods, such as blocking the service, putting harmful software, getting unauthorized access, and changing the data. These cyber-attacks can cause a lot of problems for ground-based satellite operations, such as losing the service, having wrong data, and even damaging the satellites. However, some people might think that these cyber risks are not so serious and that there are ways to prevent them or deal with them.

This literature review aims to address this gap by examining existing research, frameworks, and approaches related to evaluating cybersecurity risks in ground-based satellite operations. By consolidating and analysing the available knowledge, this review seeks to develop a framework that encompasses the critical components necessary for a thorough evaluation of cybersecurity risks and threats specific to ground-based satellite operations. The framework will consider factors such as asset identification and prioritization, risk assessment methodologies, incident detection and response mechanisms, and the legal and regulatory landscape.

This research intends to offer helpful insights and suggestions for improving the security posture of ground-based satellite operations in an increasingly difficult and interconnected cyber world by analysing the current literature, finding gaps, and providing a unique framework.

The core of satellite systems is made up of ground-based satellite operations, which include a variety of crucial tasks that make it possible to command, monitor, and communicate with satellites in orbit. The key component for assessing and managing the cybersecurity risks and threats connected with ground-based satellite operations is understanding the many elements, functions, and operational environment of these activities. By considering these elements, an efficient cybersecurity framework could be created that is specific to the needs and difficulties of ground-based satellite operations. The definition, kinds, important components, functions, and operating environment of ground-based satellite operations are all covered in this section.

Definition and types of ground-based satellite operations Ground-based satellite operations refer to the set of activities and infrastructure involved in managing and supporting satellite systems from Earth. These operations are essential for establishing communication links, collecting, and processing data, and ensuring the overall functionality and performance of satellites. Ground-based operations can be classified into several types based on their primary functions:

- Telemetry, Tracking, and Command (TT&C): TT&C operations involve monitoring and controlling satellites' health, position, and orbital parameters. They include tasks such as sending commands to adjust satellite configurations, tracking their movements, and receiving telemetry data on satellite status. One of the challenges of satellite operations is the commissioning phase, which involves validating the health and functionality of the satellite subsystems after launch. According to M. Manulis et al. (2020), commissioning typically takes about two months and requires constant monitoring and control by the ground segment using (TT&C) systems.

- Payload Operations: Payload operations focus on the utilization of satellite payloads, which can include tasks such as capturing and processing remote sensing imagery, transmitting telecommunications signals, or conducting scientific experiments. These operations involve coordinating data collection, processing, and dissemination activities (Kodheli et al., 2021).
- Network Operations: Network operations pertain to the management and maintenance of the ground-based communication infrastructure that facilitates satellite connectivity. This includes managing ground stations, coordinating data transmission and reception, and ensuring network availability and reliability. According to Kodheli et al. (2021), satellite communications can benefit from ML/AI techniques in various ways, such as reducing interference, optimizing radio resources and network operations, and managing large satellite constellations. He doesn't acknowledge the trade-offs and benefits of ML/AI techniques over other conventional methods for satellite communications, such as optimization algorithms, signal processing techniques, etc., which may have different advantages and disadvantages depending on the context and objectives.

Key components and functions of ground-based satellite operations Ground-based satellite operations involve several key components and perform various critical functions to support satellite systems. The following components and functions are integral to the operations:

- Ground Control Center (GCC): The GCC serves as the central hub for managing and controlling satellite operations. It houses the necessary infrastructure, hardware, and software systems to communicate with satellites, monitor their health, and execute commands. The GCC also facilitates data processing, storage, and analysis. One of the potential threats to the security of satellite systems is the unauthorized access to ground control center, which are responsible for communicating with and controlling the satellites. This could be achieved by exploiting poorly configured or vulnerable technologies, such as outdated software or weak encryption, or by phishing, which is a technique of deceiving users into revealing their credentials or installing malware (Manulis et al., 2020). This scenario is like the attacks on enterprise IT networks, which also rely on exploiting technical or human weaknesses to compromise the system (Manulis et al., 2020).
- Ground Stations: Ground stations are strategically located facilities equipped with antennas and receivers to establish communication links with satellites. They are responsible for transmitting commands, receiving telemetry data, and exchanging information between satellites and the GCC. Ground stations are often distributed geographically to ensure continuous coverage and efficient data transfer. Pavur and Martinovic (2022) assume that satellite ground stations are not different from any other terrestrial computing network, but they do not provide any evidence or justification for this claim. Moreover, they acknowledge that satellite ground

stations have some similarities to terrestrial communications systems, but they do not explain what these similarities are or how they affect the performance or security of the satellite network.

- Data Processing and Analysis Systems: These systems are responsible for receiving, processing, and analysing the data collected by satellites. They employ various algorithms, software tools, and databases to extract meaningful information from the collected data, perform quality checks, and generate actionable insights. Shared ground station networks, such as those offered by cloud-based services (e.g., Amazon Web Services), can facilitate the communication and data processing of different constellations by providing access to high performance computing resources (Kodheli et al., 2021).
- Command and Control Interfaces: Command and control interfaces enable operators to interact with satellites, send commands, and receive telemetry data. These interfaces provide a user-friendly environment to monitor satellite status, configure parameters, and execute operational tasks.

2 Literature Review

Ground-based satellite operations are conducted in a complex operational environment that encompasses technological, physical, and organizational aspects. This environment includes:

- Technological Infrastructure: Ground-based satellite operations rely on advanced technologies such as satellite communication protocols, data encryption mechanisms, network infrastructure, and software systems. The operational environment must address technological considerations to ensure secure and efficient communication and data management. Tsochev et al. (2020) claims that a mix of physical and technological measures, as well as consumer education, is necessary to prevent the attacks. However, they do not provide sufficient evidence or examples to support this assertion. How effective are these methods in different contexts and scenarios? What are the potential drawbacks or limitations of relying on them? How do they address the root causes of the attacks or the motivations of the attackers?
- Physical Infrastructure: Ground-based operations require dedicated facilities, including the GCC, ground stations, power and cooling systems, and physical security measures. The physical infrastructure must be designed to withstand environmental factors, provide redundancy, and protect against unauthorized access. According to Tosun (2021) he vulnerability of vital infrastructure to terrorist attacks poses a serious challenge to national security and social stability. Terrorists aim to cause maximum damage, disruption, and demoralization by targeting critical systems and services that support the country's economy and public welfare (Tosun, 2021).

- Organizational Structure: Ground-based satellite operations involve a structured organizational setup with specialized roles and responsibilities. This may include operators, engineers, cybersecurity professionals, and management personnel. The operational environment should promote effective coordination, communication, and collaboration among different stakeholders. According to Li and Liu (2021) a new trend in organizational management is to hire a senior data security manager or a senior manager who oversees the various aspects of the security status of organizations.

2.1 *Cybersecurity Risks and Threats in Ground-Based Satellite Operations*

Cybersecurity risks and threats pose significant challenges to the integrity and security of ground-based satellite operations. This section provides an overview of cybersecurity risks and threats, identifies, and classifies those specific to ground-based satellite operations, and presents case studies/examples of cybersecurity incidents in this domain.

The security of satellite-to-satellite communication links is a crucial aspect for the reliability and availability of satellite-based communication systems. However, most of the existing security solutions for SATCOMs focus on the ground segment or the user terminals, neglecting the potential threats and vulnerabilities that affect the space segment. According to Tedeschi et al. (2022) this is due to the difficulty of obtaining information about the communication protocols used by satellite links, which are often proprietary and confidential, and to the assumption that such links are inherently secure and isolated from external attacks. However, this assumption is questionable, as satellite links are exposed to various types of attacks, such as jamming, spoofing, eavesdropping, and hijacking, which can compromise the integrity, confidentiality, and availability of the transmitted data. Moreover, the authors argue that most of the current satellite systems do not implement adequate security mechanisms or rely on outdated and ineffective techniques. Therefore, they propose a novel security framework for satellite-to-satellite communication links, based on quantum key distribution and physical layer security techniques. Cybersecurity risks and threats encompass a wide range of malicious activities that can exploit vulnerabilities in the information technology infrastructure supporting ground-based satellite operations. These risks and threats showed on the Table 1 can be categorized into several key areas:

Table 1 Key cybersecurity risks and threats associated with ground-based satellite operations

Cyber threat	Definition
Data corruption/ modification	The unauthorized alteration or destruction of data transmitted or stored by satellite systems
Ground system loss	The physical or logical damage or disruption of the ground stations or networks that control or communicate with satellites
Interception of data	The unauthorized access or eavesdropping of data transmitted or received by satellite systems
Jamming	The intentional or unintentional interference or noise that affects the quality or clarity of the signals transmitted or received by satellite systems
Denial of service	The intentional or unintentional overload or flooding of the satellite systems or networks with requests or traffic that prevent them from providing normal services to legitimate users
Masquerade (spoofing)	The impersonation or falsification of the identity or source of the signals or data transmitted or received by satellite systems
Replay	The capture and retransmission of signals or data by an unauthorized party to satellite systems
Software threats	The malicious code or programs that exploit the vulnerabilities or weaknesses of the software components or applications used by satellite systems
Unauthorized access	The unauthorized entry or use of the satellite systems or networks by an unauthorized party

2.2 Identification and Classification of Cyber Threats Specific to Ground-Based Satellite Operations.

Ground-based satellite operations face unique cyber threats due to their specific operational characteristics and requirements. It makes sense and has been thoroughly researched why anyone would want to harm satellites. Identifying and classifying these threats is essential for developing targeted mitigation strategies (Livingstone & Lewis, 2016). In terms of potential attackers, a 2016 report by Chatham House, a well-known UK policy think-tank, classifies security actors into four main groups: nations looking for an edge, organized crime operations for financial gain, terrorist groups looking for recognition, and lone hackers showcasing their skills. The Consultative Committee for Space Data Systems (CCSDS)'s list of attackers may be utilised to complement this. Some specific cyber threats associated with ground-based satellite operations include:

- Command and Control System Compromises: Threats targeting the command-and-control systems used to communicate with satellites can disrupt or manipulate satellite operations. These threats may involve unauthorized command transmission, injection of malicious commands, or manipulation of telemetry data. Over the 1990s, there were several attacks on satellite ground stations. These were high-profile events when hackers allegedly gained control over systems that would have allowed them to command spacecraft in orbit. The most noteworthy of them is a situation that occurred in 1998 in which hackers, generally suspected to be connected to the Russian government, obtained access to flight control systems at NASA's Goddard Space Flight Centre (Fritz, 2013) Some others link these incursions to a subsequent event where the German-US ROSAT X-ray telescope changed its orientation to aim optical sensors straight at the sun, resulting in catastrophic hardware damage (Epstein & Elgin, 2008).

- Ground Station Vulnerabilities: Ground stations, being critical communication links with satellites, can be targeted through cyber threats. Vulnerabilities in ground station equipment, software, or network infrastructure can lead to unauthorized access, data interception, or disruption of communication. In 2016, GNSS equipment revenue accounted for $84.6 billion of the $113.4 billion in overall ground equipment revenue, which has been increasing steadily since 2012 (Bryce Space and Technology, 2017, p. 5). If a physical attack is successful in exploiting a vulnerability, the ground station may be rendered inoperable, which would have an immediate impact on how the mission is carried out and the services offered. If it wants to seize control of the spaceship without really harming the systems, it can potentially try to capture the facility (Martin, 2012). According to a NASA assessment, the theft of an unencrypted notebook computer resulted in the loss of command-and-control algorithms for the ISS (Manulis et al., 2020).
- Supply Chain Attacks: Ground-based satellite operations rely on a complex supply chain involving hardware, software, and services from various vendors. Cyber threats can exploit vulnerabilities in the supply chain, leading to compromised components, malicious software injections, or unauthorized access to sensitive systems. Authentic components could not be accessible due to supply chain disruptions, which could lead to the usage of counterfeit parts (CCSDS, 2022). If confidence is not established, fake equipment or software may be given and utilized covertly on a mission. Malicious circuits or malicious code might be included in the hardware or software, which could have unwanted mission effects. Sensitive materials may be exposed or made accessible by supply chain partners. Suppliers who are connected and have security flaws might be exploited and utilized to harm organizations, systems, and missions (CCSDS, 2022).
- Insider Threats: The involvement of personnel with privileged access to ground-based satellite operations introduces the risk of insider threats. These threats may involve intentional or unintentional actions by employees or contractors that compromise system security, leak sensitive information, or disrupt operations. The most frequent subcategories of threat actors, according to Tsochev et al. (2020), include terrorist organisations, insiders of businesses, and cybercriminals. Since it has been shown that each set of agents employs nearly similar attack tools, identifying them makes sense from a practical standpoint.

2.3 Case Studies/Examples of Cybersecurity Incidents in Ground-Based Satellite Operations

Galileo Navigation System Incident: One of the most significant incidents that affected the Galileo satellite navigation system was the outage that occurred in July 2019. According to the European Global Navigation Satellite Systems Agency (GSA), which is responsible for operating the system, the outage was caused by a technical incident related to its ground infrastructure. The incident specifically affected the European spacecraft ground station in Fucino, Italy, where atomic clocks are used to synchronise the signals from the satellites. The outage disrupted the initial navigation and timing services of Galileo, except for the search and rescue service. The outage lasted for six days and was resolved by experts who analysed the root cause and implemented recovery actions. The outage had significant impacts on various sectors and users who rely on Galileo for positioning, navigation, and timing services. For example, some airlines, railways, maritime operators, and emergency services reported disruptions or delays in their operations due to the loss of Galileo signals. Some smartphone users also experienced problems with location-based apps or services that use Galileo as a primary or secondary source of positioning data. The outage also affected some scientific experiments and projects that depend on high-precision measurements from Galileo.

The outage exposed the vulnerability of the Galileo system to technical failures and human errors, as well as the need for more robust backup and contingency plans. The outage also raised some questions and challenges for the future development and governance of the Galileo system. For example, some analysts and experts criticized the lack of transparency and communication from the European authorities during and after the outage and called for more accountability and oversight of the system. Some also suggested that the outage could undermine the trust and confidence of the users and stakeholders in the reliability and performance of Galileo, especially compared to other GNSS systems like GPS or BeiDou. Moreover, some also pointed out that the outage could have political and strategic implications for the EU, especially considering Brexit, the ongoing budget negotiations, and the creation of a new EU space agency.

The cyberattack against Viasat in January 2022 was a clear example of how space assets can be targeted in geopolitical conflicts. The attack not only disrupted the satellite broadband service for many users, but also potentially compromised the security and sovereignty of Ukraine, which was facing a military invasion from Russia. The attack also exposed the vulnerability of the ground segment of satellite operations, which relies on network devices and protocols that can be exploited by hackers. The attack highlighted the need for more robust cybersecurity measures and standards for commercial satellite operators, as well as more cooperation and coordination among stakeholders in the space domain. A possible way to improve the cybersecurity of the ground segment is to adopt a zero-trust approach, which assumes that any network element or user can be compromised and requires continuous verification and encryption of data and commands (NIST, 2021).

A major incident in 2008 that brought attention to the susceptibility of space assets to unauthorized access was the hacking of the Terra EOS AM-1 satellite (BBC News, 2010). The NASA Earth Observing System (EOS) program, which aspires to collect information on the Earth's climate, atmosphere, and surface, includes the Terra EOS AM-1 satellite. Hackers could get unauthorized access to the satellite's command and control systems during the event (BBC News, 2010). They could briefly impair the satellite's functioning thanks to this access, but there was no long-term harm or disruption. The incident did, however, serve as a reminder to pay attention to the security of space assets, especially those used for scientific and environmental study (BBC News, 2010). The incident's primary concerns included the security of satellite communications. It soon became clear that satellite communications encryption and authentication procedures were insufficiently reliable to prevent unauthorized access. More stringent authentication and encryption procedures for satellite communications might be implemented to solve this. The systems of the satellite would be harder for unauthorized users to gain entry to, and the integrity and confidentiality of the data being transmitted would be safeguarded by effective encryption. The event also brought to light the difficulties with attribution and deterrence in the online environment. The identity and intent of the hackers were never unequivocally established despite efforts to examine the hacking event. It is challenging to identify those responsible and act as a deterrence for further assaults because of this lack of attribution. Increasing technological and diplomatic efforts to detect and attribute cyberattacks will assist establish responsibility and possibly deter criminals from attacking space assets (NIST, 2021). In conclusion, the 2008 hacking of the Terra EOS AM-1 satellite highlighted the requirement for better security protocols for space assets. Enhancing attribution capabilities, monitoring, and auditing the operations of the ground segment, strengthening authentication and encryption protocols for satellite communications, and encouraging international cooperation in space security can all help to prevent similar incidents in the future and protect scientific and environmental satellite programmes (NIST, 2021).

2.4 Communication in Ground-Based Satellite Operations

Modern ground-based satellite operations are fundamentally dependent on satellite communication, which enables the transmission of data, audio, and video across immense distances with incomparable speed and worldwide coverage. It is crucial for many applications, including telecommunications, broadcasting, weather monitoring, navigation systems, scientific research, and national security activities. The way we connect, interact, and gain data on a global scale has been revolutionized by satellite communication, which uses an enormous number of geostationary and non-geostationary satellites surrounding the Earth. Some examples of satellite communication systems are:

- Global Positioning System (GPS): A constellation of 24 satellites that provides accurate location and time information to users worldwide.

Iridium: A network of 66 low-Earth orbit satellites that offers global voice and data services for mobile devices.

- Inmarsat: A provider of global mobile satellite communications for maritime, aviation, government, and enterprise sectors.
- Intelsat: A leading operator of geostationary satellites that delivers high- performance connectivity for media, broadband, government, and mobility customers.

These systems offer various benefits, such as:

- Reliability: Satellite communication is less prone to interference, congestion, or outages than terrestrial networks.
- Flexibility: Satellite communication can be deployed quickly and easily to remote or inaccessible areas where terrestrial infrastructure is lacking or damaged.
- Scalability: Satellite communication can accommodate increasing bandwidth demands and support multiple users and applications simultaneously.
- Security: Satellite communication can provide secure and encrypted transmission of sensitive data and information for military and government purposes.

To ensure smooth data transfer, satellite communication systems are made up of both space-based and ground-based components. Satellite communication system's basic components usually consist of one or more satellites placed in specified orbits, communication payloads on board, ground stations outfitted with transceivers for uplink and downlink communication, and end-user terminals or devices. The communication process begins with the transmission of data from ground-based stations to the orbiting satellites through uplink signals. These satellites then retransmit the received signals back to Earth in the form of downlink signals, which are picked up by ground stations strategically located around the globe. These ground stations serve as gateways between the satellite network and terrestrial communication infrastructure, allowing the exchange of information between satellite networks and conventional wired or wireless networks.

According to Kodheli et al. (2021) the role of satellite communication in ground-based satellite operations cannot be overstated. It plays a fundamental role in ensuring the effective operation and management of satellites, providing critical command and control capabilities for satellite positioning, orientation, and performance monitoring. Moreover, satellite communication is indispensable for transmitting telemetry and payload data, conducting software updates, and executing various operational tasks essential for satellite functionality and longevity. Furthermore, satellite communication serves as a lifeline for mission-critical applications, such as military and defence operations, emergency response, and disaster management. In remote or inaccessible regions where terrestrial communication infrastructure is limited or absent, satellite communication acts as the primary means of communication, bridging vast communication gaps and connecting people globally.

According to Pavur and Martinovic (2022) the first artificial satellites were launched into orbit in the late 1950s and early 1960s, beginning the history of satellite communication. Significant gains in satellite communication capabilities have been made over time because of developments in digital signal processing, space technology, and satellite architecture. At first, communication satellites could only enable speech and slow-speed data transmission. However, satellite communication systems have seen a tremendous metamorphosis because of the introduction of powerful satellites, several frequency bands, and advanced signal modulation techniques. Satellite systems of today provide multimedia capabilities, high-speed data transfer, and a variety of uses that were previously unthinkable.

2.5 Satellite Communication Protocols and Standards

Satellite communication protocols and standards play a crucial role in ensuring efficient and secure data exchange between ground-based stations and orbiting satellites. These protocols define the rules and conventions for data formatting, transmission, error detection, and correction, while standards provide a framework for interoperability and compatibility among different satellite communication systems. Some of the common communication protocols used in Ground-Based Satellite operations are listed below:

- CCSDS Protocols (Consultative Committee for Space Data Systems): CCSDS (2022) has developed a set of standardized protocols widely used in space missions, including satellite communications. Among these protocols, the CCSDS Telemetry, Tracking, and Command (TM/TC) protocol is particularly critical for exchanging telemetry data and commanding satellite operations. CCSDS File Delivery Protocol (CFDP) is another significant protocol used for reliable file transfer between space systems.
- SCPS Protocols (Space Communications Protocol Standards): The SCPS protocols provide a suite of optimized communication protocols tailored for satellite networks. These protocols aim to enhance the efficiency of data transmission over long-delay communication links, making them ideal for space-based operations.
- DVB-S2 (Digital Video Broadcasting—Satellite—Second Generation): DVB-S2 is a widely adopted standard for digital satellite communication, especially for broadcasting multimedia content. It offers improved spectral efficiency, higher data rates, and better error correction compared to its predecessor, DVB-S.
- TCP/IP over Satellite: Although TCP/IP is primarily designed for terrestrial networks, its adaptation for satellite communication is common. However, the long propagation delays in satellite links can lead to degraded performance. Various optimization techniques, such as TCP acceleration and proxy-based approaches, have been employed to address these challenges. As technology continues to evolve, satellite communication protocols and standards will also undergo refinements and updates to meet the growing demands for higher data

rates, improved security, and seamless interoperability. Standards listed below are the one to secure the satellite communications.

- AES Encryption (Advanced Encryption Standard): AES is widely recognized as a secure encryption standard used to protect sensitive data transmitted over satellite links. It provides strong encryption algorithms that safeguard the confidentiality and integrity of data.
- IPSec (Internet Protocol Security): IPSec is a suite of protocols used to secure internet communications, including those over satellite links. By encrypting data at the IP layer, IPSec ensures secure end-to-end communication between satellite ground stations and user terminals.
- SSPA/HPA Standards (Solid State Power Amplifier/High-Power Amplifier): These standards define the requirements and performance characteristics of power amplifiers used in satellite communication systems. Ensuring compliance with these standards helps maintain reliable and interference-free satellite communications.
- Viterbi and Reed-Solomon Coding: Error correction coding schemes like Viterbi and Reed-Solomon are employed in satellite communication to combat signal degradation caused by atmospheric and space-related interference. These coding techniques enable the recovery of lost or corrupted data during transmission.

Satellite communication security is a field that is always changing due to technological improvements and the appearance of fresh dangers (Hosseinidehaj et al., 2019). Anomaly detection systems, software-defined networking for dynamic security setups, and quantum-resistant encryption are a few promising technologies that might improve the security of satellite communication against knowledgeable adversaries.

2.6 *Satellite Communication in the Age of Quantum Computing*

The provision of dependable and secure communication lines between ground stations and orbiting satellites has advanced significantly in the field of satellite communication. Numerous crucial applications, including as communications, Earth observation, and national defence, depend on these systems. However, the development of quantum computing technology poses fresh difficulties that risk upending established secure communication paradigms. This part examines the security of satellite communications as it is now and evaluates how susceptible it is to risks from quantum computing. Techniques for encryption are used in satellite communication to protect the privacy and reliability of information sent. These encryption techniques have progressed, and contemporary systems now use cutting-edge encryption techniques like RSA and ECC (Elliptic Curve Cryptography) to safeguard sensitive data according to Thomas, A. and Manuel, E. M. (2016). Although these encryption techniques have proven resistant to classical computers, the situation changes when quantum computers are involved.

Quantum computers have the potential to break widely used encryption algorithms by exploiting their inherent vulnerabilities to quantum attacks. The most notorious among these attacks is Shor's algorithm, which can efficiently factor large integers and solve the discrete logarithm problem. Quantum computers are still in their infancy and face many technical challenges before they can achieve the level of performance and reliability required to run Shor's algorithm effectively. Moreover, there are alternative encryption algorithms that are resistant to quantum attacks, such as lattice-based cryptography and hash-based signatures. These algorithms can be used to protect satellite communication data from quantum adversaries. These two mathematical problems underlie the security of many encryption algorithms, including RSA and ECC. As quantum computers advance in their capabilities, the timeframe within which they can potentially crack encryption keys shortens. What was once considered a secure communication method against classical computers may become obsolete, raising concerns about the confidentiality of satellite communication data.

Researchers and organisations are working to create quantum-resistant cryptography methods in response to the looming threat of quantum computing. These algorithms are meant to survive quantum computer assaults and preserve data security in a post- quantum environment. There is a growing understanding among those involved in satellite communication of the necessity of switching to quantum-resistant encryption techniques. For example, to protect satellite operations, the Consultative Committee for Space Data Systems (CCSDS) is working hard to build quantum-resistant communication protocols.

Although post-quantum cryptography is advancing, applying quantum-resistant security measures in satellite communication systems poses difficulties. Satellites frequently operate in resource-constrained situations and have limited computing resources. Thus, while implementing new cryptographic algorithms, it is important to consider how much energy and computing overhead would be required. Furthermore, amongst satellite operators, ground stations, and communication equipment makers, careful planning and cooperation are required to make the shift to quantum-resistant encryption. This shift may entail hardware upgrades and adjustments to operational procedures; it is not only a software update.

The current state of satellite communication security is robust against classical computing threats. However, the impending era of quantum computing introduces a new dimension of vulnerability (Kodheli et al., 2020). To ensure the continued confidentiality and integrity of satellite communication data, the satellite industry must proactively adopt quantum-resistant encryption methods and address the unique challenges associated with their implementation in satellite systems which is going to be proposed in this framework. This transition will be pivotal in safeguarding the future of satellite communication in the quantum computing era.

3 Simulation Process

This research aims to provide a framework for evaluating the risks to cybersecurity posed by ground- based satellite activities.

The research strategy employed in this study aligns with proposing the best way to encrypt the communication between two ground station satellites. This includes conducting in-depth reviewing technical documents, standards, guidelines, and analysing relevant case studies. These methods enable the capture of rich and nuanced insights into how individuals and organizations within the satellite industry make sense of cybersecurity risks and threats.

In essence, the interpretivist philosophy in this research recognizes that cybersecurity is not solely a matter of technical protocols and security measures but is deeply intertwined with the perceptions, experiences, and interpretations of those involved. By embracing this perspective, the research aims to develop an evaluation framework that is not only technically robust but also sensitive to the diverse human factors and contextual nuances that define the cybersecurity landscape in ground-based satellite operations.

The developed framework is subsequently applied and tested through the utilization of real-world case studies within the domain of ground-based satellite operations. This empirical examination serves the purpose of assessing the framework's efficacy in the identification, mitigation, and management of cybersecurity risks, threats and how to be sure that the communication is not breakable in the future. Additionally, it facilitates the identification of potential gaps or areas necessitating further refinement.

The deductive research approach chosen for this dissertation adeptly leverages existing knowledge and established theory as a robust foundation upon which to construct a practical and all-encompassing evaluation framework. It synthesizes theoretical insights with empirical validation and iterative enhancement to ensure that the framework not only remains theoretically grounded but also remains adaptable and responsive to the dynamic and multifaceted cybersecurity landscape inherent to ground-based satellite operations. The dissertation aims to deliver a valuable and practical tool for augmenting the security posture of ground-based satellite operations.

3.1 Practical Work

The purpose of this study is to look into how quantum computing may affect the safety of satellite communication networks and provide a framework that describes a set of guidelines to ensure that satellites and ground stations uses a quantum resistant and powerful encryption and digital signatures algorithms. In order to accomplish operations that are impractical or unsolvable for classical computers, quantum computing uses the quantum mechanical features of subatomic particles. Shor's method, which effectively factors big numbers and defeats the popular Rivest–Shamir–Adleman (RSA) encryption, is one of the most well-known uses of quantum computing. The security of satellite communication systems, which depend on RSA encryption to safeguard the confidentiality and integrity of data transferred between satellites and ground stations, is seriously jeopardized by this, and using Grover Search Algorithm, the time needed to break AES will be halved. The author suggest using the lattice-based encryption method NTRU, which is immune to quantum assaults and appropriate for satellite communication applications and, another algorithm, FALCON can be used which is lattice-based digital signature scheme that offers high security and efficiency. FALCON can be used to sign the messages between the satellites and the ground stations, ensuring their authenticity and integrity. FALCON has low key size, fast verification, and low bandwidth consumption, which are suitable for SATCOM. Moreover, FALCON has been selected by NIST for standardization, which indicates its reliability and compatibility. Therefore, FALCON is a viable option for securing SATCOM in the quantum era, one of the protocols that matches these criteria, have been proposed by in (Hussien et al., 2024). This section will discuss the application and assessment of OssLayer protocol in this part. Firstly it will go over the specifics of our infrastructure for simulating quantum computing, where its used Shor's algorithm and evaluated how well it worked with various RSA key sizes. The design and configuration of our model of a satellite communication network are then shown. Here, the OssLayer protocol was used, and its security and effectiveness were compared to those of the RSA method. The author conclude by summarizing and evaluating the findings of our experiments, which show the viability and efficiency of our method for increasing the security of satellite communication systems in the era of quantum computing.

3.2 *Testing RSA Encryption Using Shor's Algorithm*

This section examines the threat posed by quantum technology to the security of the RSA, one of the most popular encryption methods by looking into the possibility of obtaining the RSA encryption's secret keys using Shor's algorithm, a quantum technique that can effectively factor big numbers. The author implement Shor's algorithm on 12 qubit machine, execute it using a quantum computing environment using IBM online platforms, and evaluate its effectiveness. Also go over the results' implications for the safety of data transmission, particularly as they relate to satellite communication systems. Seeking to address the following research queries in this section: Why is Shor's algorithm effective against RSA encryption? How does it operate? How quick and dependable is Shor's algorithm in using a quantum computing simulation to crack RSA encryption? What are the difficulties and restrictions of employing a simulation of quantum computing for cryptanalysis? How can we defend against future quantum attacks on our data? The advantages and imperfections of RSA encryption in the face of quantum computing are revealed in this part, which also offers insight into the quantum world and its impact on cryptography.

To rigorously assess the vulnerability of RSA encryption in the quantum computing era, it is executed a meticulously designed experiment within IBM's Quantum Experience platform. The journey began by gaining access to IBM's quantum computing infrastructure, an indispensable resource for implementing Shor's Algorithm. The selection of a 12-qubit quantum processor was deliberate, mirroring the early stages of quantum computing development and the realistic constraints that exist today. The heart of our investigation was the implementation of Shor's Algorithm, a quantum algorithm renowned for its ability to factorize large numbers exponentially faster than classical computers. The implementation culminated in the execution of Shor's Algorithm, targeting the factorization of a selected number. Our primary focus was on the factorization of 15.

The Shor algorithm is a quantum algorithm that has the potential to dramatically impact the field of cryptography, particularly RSA encryption (Chandana et al., 2022). RSA encryption relies on the difficulty of factoring large composite numbers into their prime factors. The security of RSA encryption is based on the belief that factoring large numbers is a computationally expensive task, even for the most powerful classical computers available today. The Shor algorithm leverages the power of quantum computing by using qubits, which can exist in multiple states simultaneously (superposition) (Faujdar & Ghrera, 2015). This allows it to perform many calculations in parallel. The core of the Shor algorithm involves finding the period of a function. In the context of factoring, this function is designed to reveal the factors of a composite number when its period is determined. Shor's algorithm efficiently finds the period using quantum operations (Chandana et al., 2022). Once the period is found, the algorithm employs classical mathematical techniques like continued fractions to extract the factors from the period. This is the classical part of the algorithm, the "a" represent the random starter number, the "r".

3.3 Results and Implications

The results gleaned from our Shor's Algorithm implementation resonate profoundly in the context of encryption security and the quantum revolution. Shor's Algorithm successfully factorized the number 15 into its prime components, 3 and 5. This milestone, though seemingly innocuous, carries profound implications. The fact that 15 could be factored successfully highlights the susceptibility of popular encryption techniques like RSA in the presence of quantum computing. The security foundations of conventional encryption techniques deteriorate as quantum computers increase in qubit count and processing capacity. This experiment serves as a resounding wake-up call, emphasizing the pressing need for the adoption of post-quantum cryptographic solutions. The transition to encryption methods resilient against quantum attacks is no longer a distant goal, it's a necessity. The implications of this quantum threat reverberate far beyond this dissertation, urging organizations, institutions, and the cybersecurity community at large to take decisive action in safeguarding sensitive data in an era where quantum supremacy is no longer a distant prospect (Fig. 1).

```
starter random number 11

/tmp/ipykernel_57/462956779.py:17: DeprecationWarning: Using
ectly using `backend.run(circuits, **run_options).
  results = qasm_sim.run(assembly, memory=True).result()

possibilities found: 15  and  1


starter random number 2
possibilities found: 1  and  3


starter random number 11
possibilities found: 15  and  1


starter random number 2
possibilities found: 3  and  5


factors of 15 = 3 & 5 !
```

Fig. 1 Results of finding a factors of 15

Simulating Ground-Based Satellite Communication with OssLayer Protocol.

In this experiment, the research will use the OssLayer Protocol as a part of the simulation, which is described in (Hussien et al., 2024), the protocol have been upgraded a bit as indicated as follows:

First, the client begins the TCP handshake by sending a normal TCP SYN packet with specific settings: destination port at 0, reserved bits at 7, and protocol version field in the IP header at 99, indicating Oss Layer support. Concurrently, the client generates its own key pair using the NTRU algorithm.

Upon receiving the handshake request, the server verifies the protocol version, destination port, and reserved bits and acknowledge the OssLayer support, the then server completes the handshake generates its own NTRU key pair. The client then receives the server's public key, encrypts a random message with it, and sends it back to the server.

The server decrypts this message using its private key, obtaining a shared secret. Both parties then adopt xPoly20chacha1205 as their symmetric encryption algorithm. The server then sends a FALCON-signed message that includes an NTRU public key that is used as the server's ID, this message is encrypted with the shared secret key, to assure the client of its legitimacy, which is changed from the original haraka v2 based hash id to ensure scalability and easy deployment.

The client then connects to the PKI to verify the server's authenticity. If the server is verified, the PKI confirms this with the client, and now as of the new update is the OssLayer, the client sends a challenge message that is encrypted with this public key, to further confirms that the client is communicating with a legitimate server, not a replayed message from a man in the middle, the server have to decrypt this message not only with the shared secret, but also with it's NTRU private key that is generated as a part of it's public certificate. If the server was able to solve this challenge, it will send a challenge response message to the client, which will confirm the server identity and now the client goes through the same process, this new updates on the OssLayer protocol is shown in the Fig. 2

After the verification is complete and the mutual trust between the client and the server is achieved, the client sets the destination port for the required service and proceed with encrypting the transport layer source and destination port fields. The server decrypts the transport layer, checks the client's permissions based on its public NTRU certificate, and if the client have the required permissions to access the specified service, they can now begin the data exchange process (Fig. 3).

All algorithms used, including xPoly20chacha1205, NTRU, FALCON, and Haraka-v2, are enhanced with an experimental QRNG based on Qiskit platform for nonce generation. For each connection, the server generates two nonces as with the old OssLayer protocol, one for the server and one for the client, the client is expected to use its nonce in the next communication link (Fig. 4).

OssLayer TCP Header as described in (Hussien et al., 2024).

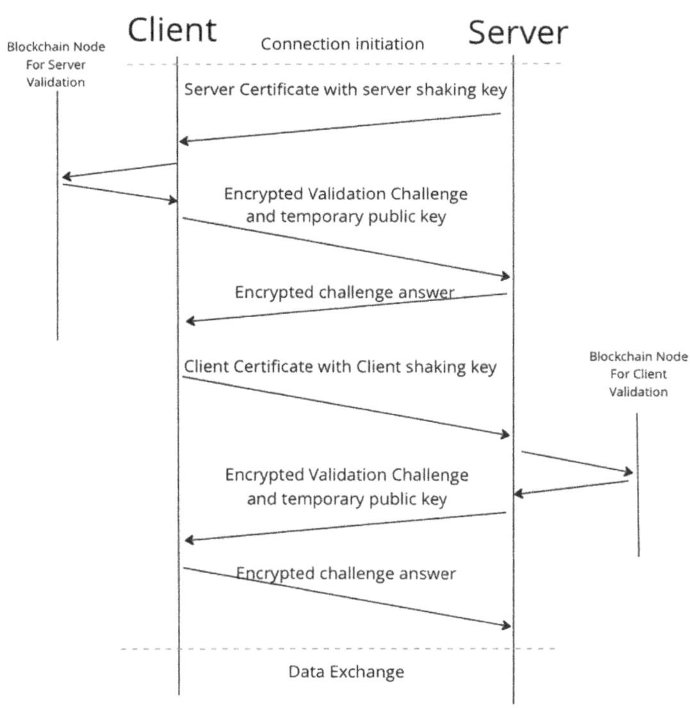

Fig. 2 OssLayer 1.1 updated handshake

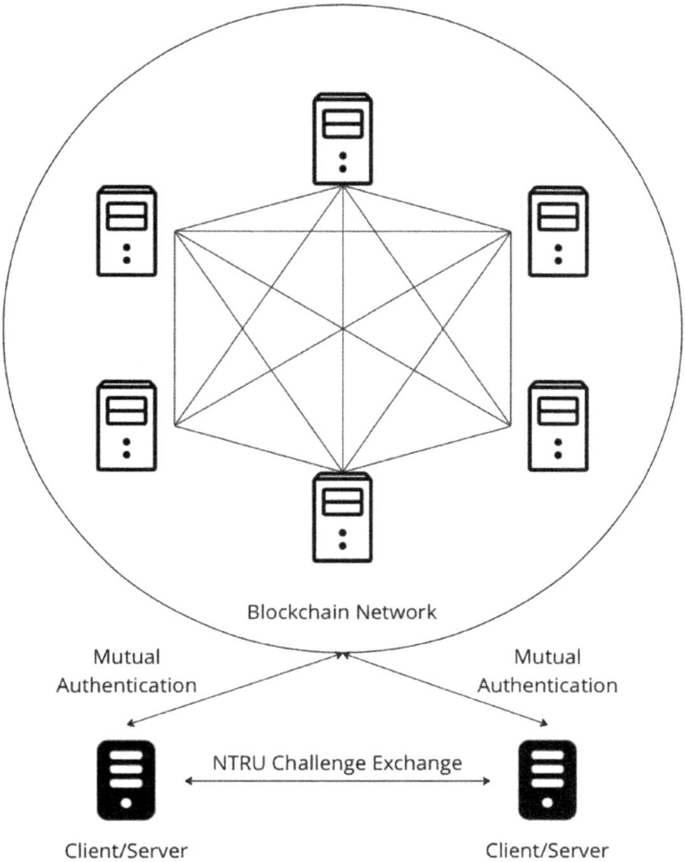

Fig. 3 Decentralized harakav2 based Blockchain PKI Infrastructure

Source port (5 bytes)								Destination port (5 bytes)
Sequence number*								
Acknowledgement number*								
Header length*	Reserved bits*	URG*	ACK*	PSH*	RST*	SYN*	FIN*	Window size* (advertisement window)
Check sum*								Urgent pointer*
Options Used for sharing public keys/Hash ID (variable bytes size)								
Data* (optional)								

*Denotes that this field was not changed from the default Transport layer

Fig. 4 Schema

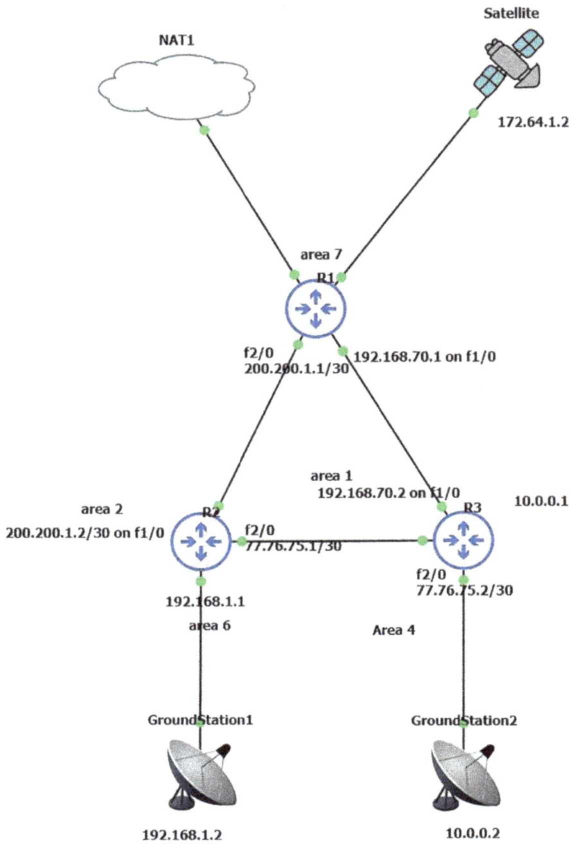

This section presents a simulation study conducted within the GNS3 network emulation environment, focusing on the communication between two client ground stations through a server satellite, utilizing the updated OssLayer Protocol. The experiment's primary objectives were to assess the performance and security of PQC cryptographies used in this experimental version of this protocol in the context of satellite communications and to evaluate its potential for real-world applications. The study involved the emulation of satellite communication scenarios, encompassing data transmission, encryption, and decryption processes. The results and insights gained from this experiment contribute valuable knowledge to the field of satellite communication security and cryptographic techniques.

3.4 Communication Simulation

The experimental setup involved the creation of a simulated satellite communication network using GNS3. Two client ground stations were established as communication endpoints, with a server satellite serving as an intermediary. Data packets were generated and transmitted between the clients via the server satellite. NTRU lattice-based cryptography was implemented to secure the communication channel. The encryption and decryption processes were integrated into the simulation to assess their impact on performance and security (Figs. 5, 6, and 7).

The first thing we need to be sure of is that both ground station satellite 1 & 2 are connected to a satellite server in order for the communication to happen (Fig. 8).

As wireshark doesn't support the OssLayer protocol, the packet details are only shown in binary format as shown below (Figs. 9 and 10).

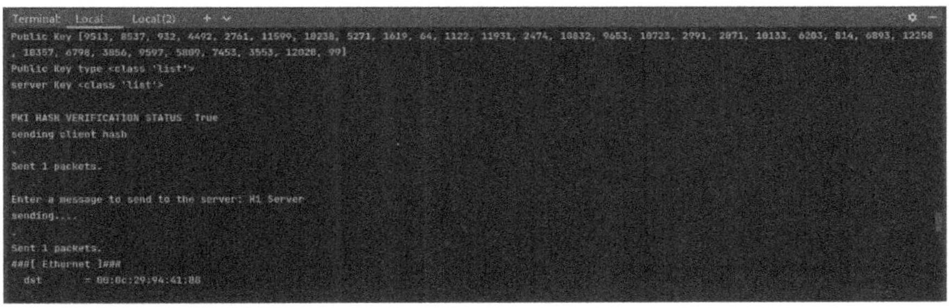

Fig. 5 Client sending "Hi Server" message to the server

Fig. 6 Server replying to the client with "Hello Client,, you said: Hi server"

Fig. 7 Server decrypting the received message from the client, verifying the client used the new QRNG nonce, and accepting the connection

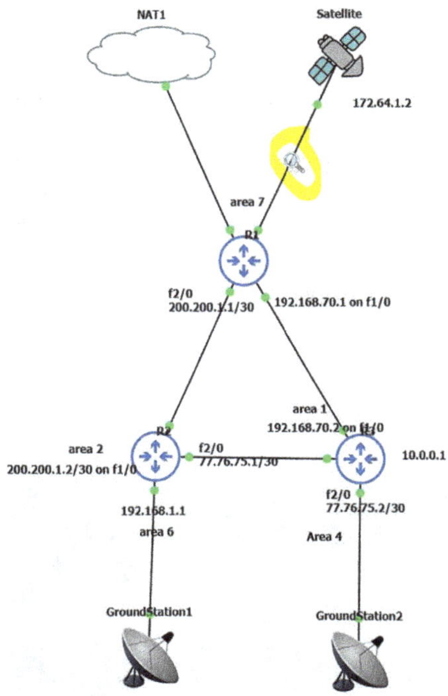

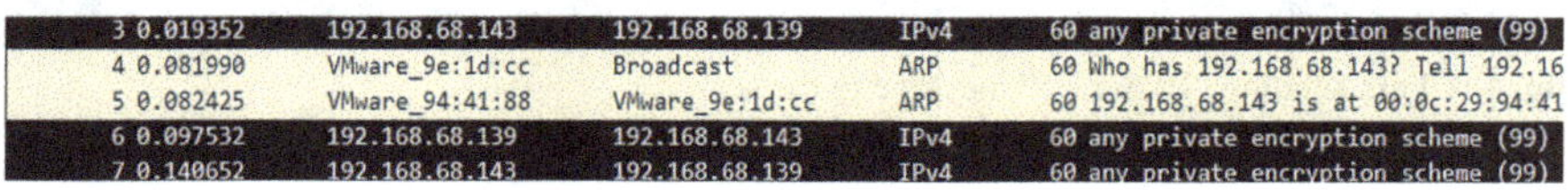

3	0.019352	192.168.68.143	192.168.68.139	IPv4	60 any private encryption scheme (99)
4	0.081990	VMware_9e:1d:cc	Broadcast	ARP	60 Who has 192.168.68.143? Tell 192.16
5	0.082425	VMware_94:41:88	VMware_9e:1d:cc	ARP	60 192.168.68.143 is at 00:0c:29:94:41
6	0.097532	192.168.68.139	192.168.68.143	IPv4	60 any private encryption scheme (99)
7	0.140652	192.168.68.143	192.168.68.139	IPv4	60 any private encryption scheme (99)

Fig. 8 OssLayer TCP Handshake packet capture

Fig. 9 Syn packet details in wireshark

0000 00 0c 29 9e 1d cc 00 0c 29 94 41 88 08 00 45 00..)…..).A…E.
0010 01 76 00 01 00 00 40 63 6e b9 c0 a8 44 8f c0 a8.v….@cn…D…
0020 44 8b 00 00 00 00 00 00 00 00 00 00 00 00 00 00 D…………..
0030 00 00 00 00 00 18 20 00 00 00 00 00 7b 22 6e 6f …… …..{“no
0040 6e 63 65 22 3a 20 22 72 72 72 72 72 72 72 72 72 nce”: “rrrrrrrrr
0050 72 72 72 72 72 72 72 22 2c 20 22 68 65 61 64 65 rrrrrrr”, “heade
0060 72 22 3a 20 22 5a 47 46 30 59 56 39 6c 65 47 4e r”: “ZGF0YV9leGN
0070 6f 59 57 35 6e 5a 56 38 74 4f 44 55 3d 22 2c 20 oYW5nZV8tODU=“,

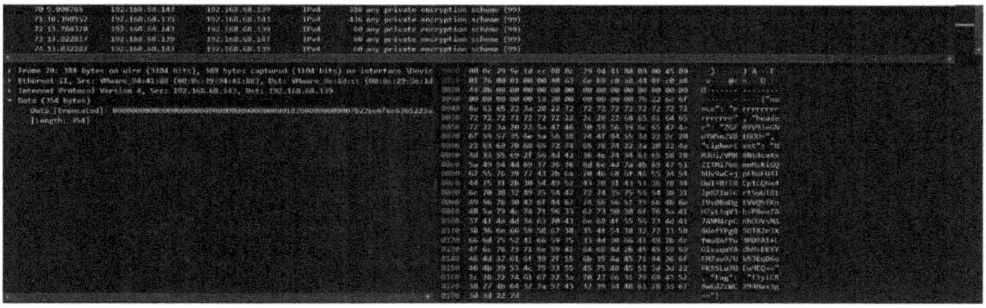

Fig. 10 The client–server message in wireshark

0080 22 63 69 70 68 65 72 74 65 78 74 22 3a 20 22 4e "ciphertext": "N
0090 4d 33 55 69 2f 56 4d 42 36 4e 74 34 63 65 58 78 M3Ui/VMB6Nt4ceXx
00a0 5a 49 54 4d 69 37 36 36 6d 6e 4d 7a 4b 69 47 51 ZITMi766mnMzKiGQ
00b0 62 55 76 39 77 43 2b 6a 70 46 68 6f 46 55 34 54 bUv9wC+jpFhoFU4T
00c0 44 75 31 2b 30 54 49 52 43 70 31 43 51 36 78 34 Du1+0TIRCp1CQ6×4
00d0 6c 70 38 32 49 75 54 47 72 74 35 75 55 54 38 31 lp82IuTGrt5uUT81
00e0 49 56 76 30 42 6f 44 67 74 56 56 51 35 66 4b 6e IVv0BoDgtVVQ5fKn
00f0 48 5a 79 4c 74 71 56 33 62 73 50 38 6f 76 5a 41 HZyLtqV3bsP8ovZA
0100 37 41 4e 4d 34 63 70 43 6e 68 4f 55 56 73 4d 41 7ANM4cpCnhOUVsMA
0110 38 36 6e 66 59 50 67 38 35 4f 54 38 32 72 33 58 86nfYPg85OT82r3X
0120 66 6d 75 52 41 66 59 75 33 4d 30 66 41 49 2b 4c fmuRAfYu3M0fAI+L
0130 4f 6c 76 73 71 6e 59 41 64 68 4d 2b 45 45 59 59 OlvsqnYAdhM+EEYY
0140 46 4d 37 61 6f 39 2f 55 6b 35 4a 45 71 44 36 6f FM7ao9/Uk5JEqD6o
0150 46 4b 39 53 4c 75 33 55 45 75 48 45 51 3d 3d 22 FK9SLu3UEuHEQ== "
0160 2c 20 22 74 61 67 22 3a 20 22 66 31 79 69 43 52, "tag": "f1yiCR
0170 38 77 4b 64 32 7a 57 43 32 39 34 48 61 78 33 67 8wKd2zWC294Hax3g
0180 3d 3d 22 7d =="}

The following showcases the packet details that are being sent, with the QRNG nonce, as of now, the QRNG is bing ran on simulation and is not production ready as can be seen from the packet details, followed by the encrypted message, and the message tag, in production environment, the QRNG will be ran directly using IBM Qiskit platform, and not only the message will be encrypted, instead the whole packet which is the nonce, the ciphertext and the tag combined.

The message sent from the client was captured using Wireshark which is using OssLayer.

The reason behind of this framework that suggest using the NTRU is to secure a communication in ground based satellites against the quantum computers with more qubits power in the future. Quantum computers can exploit the mathematical properties of large numbers and curves to break the security of these cryptosystems, but they cannot efficiently solve the problem of finding the shortest vector in a lattice, which is the basis of the Osstech algorithm which is mainly based on PQC Algorithms.

3.5 Comparative Analysis and Evaluation

In the realm of encryption security and secure communication, this research has embarked on a journey of exploration and evaluation. The author has navigated the intricate landscapes of quantum computing's impending impact on encryption through Shor's Algorithm and ventured into the world of ground-based satellite communication, seeking secure solutions in the face of evolving cyber threats. This section embarks on the final leg of our journey, wherein is conducted a comparative analysis that serves as the keystone of our investigation. Through this analysis, the author trying to illuminate the contrasting fates of conventional encryption and post-quantum cryptographic methods. Subsequently, a conclusion is written down that resonate far beyond the confines of these experiments, shedding light on the path forward in safeguarding the digital realms upon which our modern world depends.

The security of many currently used cryptography techniques is seriously threatened by quantum computers. The Advanced Encryption Standard (AES), a symmetric-key method that is based on the idea of substitution-permutation networks, is one of the most well- known and commonly utilized. AES is susceptible to Grover's algorithm quantum attack, which can drastically lower its degree of security by a factor of two. Therefore, a quantum computer can decrypt a 256-bit AES key with the same amount of work that a normal computer can decrypt a 128-bit AES key. On the other hand, there are some asymmetric key algorithms that are impervious to quantum assaults since they are based on the idea of lattice cryptography. The NTRU Algorithm, which is based on the difficulty of finding the shortest vector in a high-dimensional lattice, is one of the most well-known and effective lattice-based algorithms. NTRU has been demonstrated to be quicker than RSA for private-key operations and can be used for both encryption and digital signatures. AES, RSA and NTRU will be compared in this section based on several different factors, including key size, performance, security, flexibility, simplicity, and maturity. The key distinctions between AES, RSA and NTRU are listed in the following Table 2.

Table 2 compares six different cryptographic algorithms in terms of their key size, performance, security, flexibility, simplicity, and maturity. The algorithms are AES, NTRU, RSA, FALCON, XCHACHA20POLY1305, and BLAKE3. AES is a symmetric-key algorithm that is fast and efficient, but vulnerable to quantum attacks. NTRU and FALCON are lattice- based algorithms that are resistant to quantum attacks, but more complex and less intuitive than AES. RSA is a public-key algorithm that is based on factoring large numbers, but also vulnerable to quantum attacks. XCHACHA20POLY1305 is a stream cipher with a message authentication code that is faster than AES-GCM without hardware acceleration, but not standardized. BLAKE3 is a hash function that is faster than SHA-256 and SHA-3, but less studied and standardized than them. Table 2 shows the trade-offs between these algorithms and the factors that affect their suitability for different applications.

RSA has long been regarded as the gold standard for data security, and for good reason (Alagic et al., 2018). However, the results of our Shor's Algorithm implementation on a 12-qubit quantum computer provide a sobering glimpse into the

Table 2 Comparison of algorithms

Algorithm	Key size	Performance	Security	Flexibility	Simplicity	Maturity
AES	128–256 bits	Fast and efficient	Vulnerable to Grover's algorithm, but can be mitigated by doubling the key size	Can be combined with other algorithms such as RSA or ECC for digital signatures	Relatively simple and well-understood	Widely used and standardized
NTRU	512–2048 bits	Faster than RSA for private-key operations, but slower than AES	Resistant to quantum attacks	Can be used for both encryption and digital signatures	More complex and less intuitive than AES	Less studied and standardized than AES, but has been around for more than 20 years
RSA	1024–4096 bits	Slower than AES for both public-key and private-key operations	Vulnerable to Shor's algorithm, but can be mitigated by doubling the key size	Can be used for both encryption and digital signatures, but requires padding schemes for security selection and implementation	Based on simple mathematical problem of factoring large numbers, but requires careful parameter	One of the oldest and most widely used public-key algorithms, but facing challenges from quantum computing

(continued)

Table 2 (continued)

Algorithm	Key size	Performance	Security	Flexibility	Simplicity	Maturity
FALCON	512–1024 bits	Faster than RSA for signature generation, but slower than AES for verification	Resistant to quantum attacks	Can be used for digital signatures (based on the theoretical framework of Gentry, Peikert and Vaikuntanathan for lattice-based signature schemes)	More complex and less intuitive than AES (based on NTRU lattices with a trapdoor sampler called "fast Fourier sampling")	Less studied and standardized than AES, but has been around since 2017
XCHACHA 20POLY13 05	256 bits	Faster than AES-GCM without hardware acceleration	Resistant to nonce misuse	Can be used for encryption (takes as input a 256-bit key and a 96- bit nonce to encrypt a plaintext, with a ciphertext expansion of 128-bit)	Relatively simple and well- understood	Not standardized, but widely used in various protocols (used in IPsec, SSH, TLS 1.2, DTLS 1.2, TLS 1.3, WireGuard, S/ MIME 4.0, OTR v4 and multiple other protocols)

future of encryption (Chen et al., 2016). Successfully factorizing the number 15 into its prime components of 3 and 5 serves as a compelling reminder that the era of quantum computing is dawning, and with it, the vulnerability of traditional encryption methods. As quantum computers scale up in qubit count and processing power, RSA's invincibility wanes. This experiment is a clarion call for organizations and institutions to expedite the adoption of post- quantum cryptographic solutions. The future security of sensitive data hinges on this imperative shift.

Table 2 also indicates that RSA is based on a simple mathematical problem of factoring large numbers, which is easy to understand and explain. This is similar to AES, which is based on a simple concept of substitution-permutation networks. However, RSA requires careful parameter selection and implementation to avoid errors and vulnerabilities (Boneh, 1999). This makes RSA more challenging to use correctly than AES. Table 2 finally shows that RSA is one of the oldest and most widely used public-key algorithms, which means that it has a high level of maturity and trust (Marican et al., 2022). This is similar to AES, which is also widely used and standardized. However, RSA is facing challenges from quantum computing, which may render it obsolete in the future (Chen et al., 2016). This makes RSA less future-proof than NTRU, which is designed to resist quantum attacks (Albrecht et al., 2019).

However, NTRU is protected against quantum attacks like Shor's algorithm, which can defeat most asymmetric-key algorithms like RSA or ECC (Albrecht et al., 2019). As a result, NTRU can provide more protection against quantum computers than AES and RSA (Nguyen et al., 2023). The ability of NTRU combined with F to execute encryption and digital signatures using the same algorithm gives it an edge over AES in terms of usefulness and adaptability (Albrecht et al., 2019). However, NTRU is less intuitive and more sophisticated than AES, making it potentially more difficult to comprehend and accurately apply. NTRU is less researched and standardised than AES, hence it could have certain flaws or vulnerabilities that are unrecognised or unsolved (Albrecht et al., 2019).

Within the context of simulated ground-based satellite communication, our implementation of the NTRU lattice-based algorithm paints an optimistic picture of cryptographic resilience. Traditional encryption methods have often struggled to balance security and efficiency, requiring substantial computational resources. In stark contrast, the NTRU algorithm, designed on the foundations of lattice-based cryptography, demonstrated robust security without compromising performance. Measurements of latency and throughput indicated that NTRU cryptography meets the rigorous demands of real-world satellite communication. This comparative evaluation underscores NTRU's readiness as a pragmatic choice for securing ground- based satellite communication networks. Its resilience to quantum attacks and compatibility with operational requirements mark it as a beacon of hope in the age of quantum uncertainty.

Satellite communication (SATCOM) requires quantum-resistant cryptographic algorithms to protect the confidentiality and integrity of the messages exchanged between the satellites and the ground stations. NTRU is a lattice-based algorithms that are based on the hardness of finding short vectors in high-dimensional lattices, which is mainly used for generating shared secret, while FALCON is a digital signature scheme that can sign and verify the messages. XChaCha20-Poly1305 is an encryption algorithm that uses the shared secret generated from the NTRU key exchange and encrypt the messages. These algorithms are used in the OssLayer protocol in (Hussien et al., 2024) and can provides both a robust confidentiality and integrity for SATCOM, as well as quantum resistancy.

Table 2 helps us understand the differences between AES, RSA and NTRU algorithms based on various criteria. Table 2 shows that each algorithm has its own strengths and weaknesses, but it makes clear that regarding the danger from quantum computers with more qubits in the future, NTRU looks to be more resistant and much more complex to be breakable.

Findings and Analysis

In this section, the author present the key findings of the research and analyse these findings in the context of the research objectives and hypotheses. The findings are organized based on the research questions and objectives outlined in the earlier sections.

Ground-based satellite operations play a pivotal role in the management, control, and communication with satellites orbiting the Earth. These operations are essential for ensuring the functionality and longevity of various types of satellites, including communication, Earth observation, and scientific satellites. Ground-based satellite operations involve a complex infrastructure consisting of multiple components and functions. These components can be categorized as follows:

- Ground Stations
- Satellite Operations Centres
- Telemetry, Tracking, and Command (TT&C)
- Data Processing and Analysis
- Security Protocols

Our findings emphasize that ground-based satellite operations are a critical backbone of satellite missions, ensuring that satellites function optimally and deliver on their intended objectives.

In the practical work the author conducted a series of simulations to assess the feasibility and security of NTRU lattice-based encryption in the context of ground-based satellite communication. Our findings include:

Security Strength: NTRU encryption demonstrated resilience against quantum attacks, making it a viable option for securing satellite communication channels in the quantum era.

Computational Efficiency: The computational overhead of NTRU encryption was within acceptable limits for real-time satellite communication. This suggests that NTRU can be implemented without significant performance degradation.

Quantum-Resistant Alternative: NTRU lattice-based encryption emerged as a promising quantum-resistant alternative to traditional encryption methods for satellite communication.

Through practical testing of RSA encryption using Shor's algorithm, we confirmed the potential vulnerability of classical RSA encryption to quantum computing. Our findings underscore the urgent need for organizations to reevaluate their encryption strategies and consider transitioning to quantum-resistant algorithms, especially for sensitive data. Our simulations of ground-based satellite communication using the NTRU lattice-based encryption algorithm have revealed promising results. NTRU encryption demonstrated both robust security against quantum attacks and computational efficiency, making it a viable choice for securing satellite communication channels in the quantum era.

The implications of our research extend to various facets of ground-based satellite operations, cybersecurity, and encryption practices:

Operational Resilience: Organizations and agencies operating satellites must invest in enhancing the resilience of ground-based operations. This includes improving security protocols, redundancy measures, and training for personnel.

Quantum-Ready Infrastructure: As quantum computing advances, ground-based operations should prepare for the quantum threat. This entails assessing and upgrading infrastructure to withstand quantum attacks on communication and control systems.

Transition to Quantum-Resistant Algorithms: Our findings strongly advocate for the transition from classical RSA encryption to quantum-resistant algorithms, especially for critical data and communication channels. This transition should be a priority for organizations in various sectors.

Evaluation of NTRU Encryption: Organizations involved in satellite communication should consider the adoption of NTRU lattice-based encryption. Its quantum resistance and computational efficiency make it an attractive choice for securing satellite communication.

In conclusion, our research has shed light on the intricate interplay between ground- based satellite operations, cybersecurity, and quantum computing. The quantum threat to classical encryption methods is imminent, and the adaptation of quantum-resistant strategies is imperative. Moreover, NTRU lattice-based encryption offers a promising avenue for securing satellite communication in this evolving landscape.

In the ever-evolving landscapes of cybersecurity and satellite communication, this research journey has traversed critical terrain related to encryption security and secure data transmission. The insights gleaned from our implementation scenarios reverberate across the digital realm, carrying profound implications for the safeguarding of information.

The vulnerabilities laid bare in RSA encryption through Shor's Algorithm serve as an unequivocal signal that the clock is ticking. The inexorable march of quantum computing capabilities poses an existential threat to traditional encryption standards. The urgency is palpable; organizations and institutions must not be complacent. The transition to post- quantum cryptography is no longer a mere option but an imperious necessity. The resilience of our digital infrastructure depends on this pivotal step.

Conversely, the successful implementation of the Osslayer protocol which is based on NTRU lattice-based algorithm for shared secret generation, as well as the public key in the PKI Infrastructure, FALCON as a digital signature and XCHACHA20-Poly1305 as symmetric key within a ground-based satellite communication network illuminates a path forward on quantum resistant communication. NTRU cryptography isn't just a solution; it's a paradigm shift. Its robust security guarantees, combined with commendable performance metrics, position it as a beacon of hope in the looming quantum era. It is a testament to human innovation's ability to counter emerging threats with foresight and adaptability.

In conclusion, the dual implementation scenarios in this research have yielded invaluable insights into encryption security and secure communication's metamorphosis in the face of evolving technological threats. As quantum computing advances, the adoption of post- quantum encryption ceases to be a matter of choice. It is the cornerstone of our digital resilience, the safeguard against the quantum storm. Stakeholders across the cybersecurity and satellite communication domains must heed these findings, charting a course that ensures the integrity, confidentiality, and security of our digital future. The hour is late, but the possibilities are boundless, for in the crucible of challenge, human innovation thrives.

Our investigation into the impact of quantum computing on traditional encryption schemes, demonstrated through testing RSA encryption using Shor's algorithm, has revealed vulnerabilities in current encryption methods. The results emphasize the urgent need to transition towards post-quantum cryptography in ground-based satellite communication. Our simulation of ground-based satellite communication using the updated OssLayer protocol yielded promising results, showcasing robust security potential in the era of quantum computing. Performance evaluations indicate the feasibility of implementing this encryption scheme in real-world satellite communication systems. Additionally, a comparative analysis between traditional RSA encryption and the NTRU lattice-based algorithm underscores the necessity of adopting quantum-resistant encryption methods for ground-based satellite operations. Notably, the NTRU lattice-based algorithm exhibited superior resilience against quantum attacks.

Based on our findings, we strongly recommend that organizations involved in ground- based satellite operations initiate a phased transition to quantum-resistant cryptographic protocols. This transition should prioritize the implementation of encryption schemes like the NTRU lattice-based algorithm, which have demonstrated resistance to quantum attacks. As the field of quantum computing evolves, it is imperative for stakeholders to remain vigilant and adaptable. Continuous monitoring of advancements in quantum computing and their implications for cryptography is essential. Organizations should be prepared to update their security measures accordingly. To address the evolving cybersecurity landscape, collaboration and information sharing among satellite operators, government agencies, and cybersecurity experts are vital. Establishing industry-wide best practices and standards for quantum-resistant encryption is crucial for the long-term security of ground-based satellite operations.

4 Conclusion and Future Work

In this dissertation, we have undertaken a comprehensive exploration of the challenges posed by quantum computing to ground-based satellite operations, with a specific focus on cybersecurity. The research has showed several significant findings and insights:

The investigation into quantum computing's impact on encryption schemes has highlighted the vulnerability of traditional methods, exemplified by the successful testing of RSA encryption using Shor's algorithm. This unequivocally demonstrates the urgent need for transitioning to post-quantum cryptography in the realm of ground-based satellite communication.

The simulation of ground-based satellite communication using the NTRU lattice- based algorithm combined with Falcon and XCHACHA20-Poly1305 has presented a promising alternative. Our performance evaluation indicates that this encryption scheme holds substantial potential for ensuring the security of satellite communication systems in the era of quantum computing.

The comparative analysis between RSA encryption and the NTRU lattice-based algorithm reaffirms the pressing need to embrace quantum-resistant cryptographic protocols. The NTRU lattice-based algorithm, with its resilience against quantum attacks, emerges as a robust choice for safeguarding ground- based satellite operations.

While this research has made significant strides in understanding and addressing the cybersecurity challenges posed by quantum computing, several avenues for future research and exploration remain open. Some of them that might be consider listed as below:

Future research should delve deeper into the development and refinement of quantum-resistant encryption protocols tailored explicitly for satellite communication. Investigating alternative algorithms and their performance in satellite environments will be crucial.

The security of satellite networks, including secure key distribution and network-level security, warrants further investigation. As ground-based satellite operations become increasingly interconnected, comprehensive security measures are essential.

The impact of emerging technologies, such as quantum key distribution (QKD), on ground-based satellite operations merits exploration. Assessing the feasibility and integration of QKD into satellite communication systems is a promising avenue for research.

The development of regulatory and policy frameworks to ensure compliance with evolving cybersecurity standards in ground-based satellite operations. Collaboration between regulatory bodies, industry stakeholders, and cybersecurity experts is vital.

Needs to be consider also some limitations which are listed below:

Ensuring access to a more than 12 qubits in quantum computing environment to provide more robust evaluation.

Gathering real-world data on cybersecurity incidents in ground-based satellite operations.

Managing time effectively to meet research and writing deadlines.

Adapting to the evolving landscape of quantum computing and cryptography.

This dissertation serves as a foundational step toward securing ground-based satellite operations in the face of quantum computing advancements. By implementing quantum- resistant encryption, fostering collaboration within the industry, and pursuing further research in emerging technologies, the field can navigate the quantum era with confidence, ensuring the confidentiality and integrity of satellite communication systems.'

References

Alagic, G., Alperin-Sheriff, J., Apon, D., Cooper, C., Dang, Q., Miller, C., Moody, D., Peralta, R., Perlner, R., Robinson, A., Smith-Tone, D., & Liu, Y.-K. (2018). Status Report on the First Round of the NIST Post-Quantum Cryptography Standardization Process.

Albrecht, M. R., Curtis-Maury, M., Davidson, A., Player, R., & Postlethwaite, E. W. (2019). Estimate All the {LWE, NTRU} Schemes!.

BBC News. (2010). *Hackers 'Seized Control' of US Satellites*. https://www.bbc.com/news/technology-11504328

Bryce Space and Technology: State of the Satellite Industry Report 2017. Resreport (2017). https://brycetech.com/downloads/SIA_SSIR_2017.pdf

Boneh, D. (1999). Twenty Years of Attacks on the RSA Cryptosystem. *Notices of the AMS, 46*(2), 203–213.

Chandana, V., Tejaswini, G., Gupta, L., & Yadav, R. N. (2022). Algorithm Development Analysis: Searching and Sorting. [online] IEEE Xplore. https://doi.org/10.1109/SCEECS54111.2022.9740936

Chen, L., Jordan, S., Liu, Y. K., Moody, D., Peralta, R., Perlner, R., & Smith-Tone, D. (2016). *Report on Post-Quantum Cryptography (No. NISTIR-8105)*. National Institute of Standards and Technology (NIST).

deloitteeditor. (n.d.). *Quantum Technology Comes of Age*. [online] WSJ. https://deloitte.wsj.com/articles/quantum-technology-comes-of-age-01622574129. Accessed 19 August 2023.

Epstein, K., & Elgin, B. (2008). Network Security Breaches Plague NASA Repeated Attacks from Abroad on NASA Computers and Web Sites Are Causing Consternation Among Officials and Stirring National Security Concerns Space Shuttle Discovery Preparing for Launch in July 2005 NASA/SSPL/The Image Works. [online] https://www.cs.clemson.edu/course/cpsc420/material/Papers/NASA.pdf

Faujdar, N., & Ghrera, S. P. (2015). *A Detailed Experimental Analysis of Library Sort Algorithm*. [online] IEEE Xplore. https://doi.org/10.1109/INDICON.2015.7443165

Fritz J. Satellite Hacking: A Guide for the Perplexed. *Culture Mandala, 10*(1), 5906.

Hosseinidehaj, N., Babar, Z., Malaney, R., Ng, S. X., & Hanzo, L. (2019). Satellite-Based Continuous-Variable Quantum Communications: State-of-the-Art and a Predictive Outlook. *IEEE Communications Surveys & Tutorials,* [online] *21*(1), 881–919. https://doi.org/10.1109/COMST.2018.2864557

Kodheli, O., Lagunas, E., Maturo, N., Sharma, S. K., Shankar, B., Montoya, J. F. M., Duncan, J. C. M., Spano, D., Chatzinotas, S., Kisseleff, S., Querol, J., Lei, L., Vu, T. X., & Goussetis, G. (2021). Satellite Communications in the New Space Era: A Survey and Future Challenges. *IEEE Communications Surveys & Tutorials,* [online] *23*(1), 70–109. https://doi.org/10.1109/comst.2020.3028247

Lightman, S., Suloway, T., & Brule, J. (n.d.). NIST Interagency Report NIST IR 8401 Satellite Ground Segment Applying the Cybersecurity Framework to Satellite Command and Control. [online] https://doi.org/10.6028/NIST.IR.8401

Li, Y., & Liu, Q. (2021). A Comprehensive Review Study of Cyber-Attacks and Cyber Security; Emerging Trends and Recent Developments. *Energy Reports, 7*(7), 8176–8186. https://doi.org/10.1016/j.egyr.2021.08.126

Livingstone, D., & Lewis, P. (2016). *Space, the Final Frontier for Cybersecurity?* [online] https://www.chathamhouse.org/sites/default/files/publications/research/2016-09-22-space-final-frontier-cybersecurity-livingstone-lewis.pdf. Accessed 27 September 2020.

Manulis, M., Bridges, C. P., Harrison, R., Sekar, V., & Davis, A. (2020). Cyber Security in New Space. *International Journal of Information Security*. https://doi.org/10.1007/s10207-020-00503-w

Hussien, O. A. A. M., Arachchige, I. S. W., Jahankhani, H. (2024). Strengthening Security Mechanisms of Satellites and UAVs Against Possible Attacks from Quantum Computers. In H. Jahankhani (Eds.), *Cybersecurity Challenges in the Age of AI, Space Communications and Cyborgs. ICGS3 2023. Advanced Sciences and Technologies for Security Applications*. Springer, Cham. https://doi.org/10.1007/978-3-031-47594-8_1

Marican, M. N. Y., Razak, S. A., Selamat, A., & Othman, S. H. (2022). Cyber Security Maturity Assessment Framework for Technology Startups: A Systematic Literature Review. *IEEE Access, 11*, 1–1. https://doi.org/10.1109/access.2022.3229766

NASA Cybersecurity: An Examination of the Agency's Information Security Statement of Paul K. Martin Inspector General National Aeronautics and Space Administration. (2012). http://www.csri.info/wp-content/uploads/2012/08/HHRG-112-SY21-WState-PMartin-20120229-1.pdf

NIST. (2021). *Introduction to Cybersecurity for Commercial Satellite Operations: Draft NISTIR 8270 Available for Comment*. https://www.nist.gov/news-events/news/2021/06/introduction-cybersecurity-commercial-satellite-operations-draft-nistir

Nguyen, T. D., Nguyen, T. T., Nguyen, T. H., Tran, X. N. & Nguyen, D. T. (2023). A Hardware Architecture of NIST Lightweight Cryptography Applied in IPSec to Secure High-Throughput Low-Latency IoT Networks. *IEEE Access*.

Pavur, J., & Martinovic, I. (2022). Building a Launchpad for Satellite Cyber-Security Research: Lessons from 60 Years of Spaceflight. *Journal of Cybersecurity,* [online] *8*(1). https://doi.org/10.1093/cybsec/tyac008

Report Concerning Space Data System Standards SECURITY THREATS AGAINST SPACE MISSIONS INFORMATIONAL REPORT CCSDS 350.1-G-3 GREEN BOOK. (2022). https://public.ccsds.org/Pubs/350x1g3.pdf

Tedeschi, P., Sciancalepore, S., & Di Pietro, R. (2022). Satellite-Based Communications Security: A Survey Of Threats, Solutions, and Research Challenges. *Computer Networks, 216*, 109246. https://doi.org/10.1016/j.comnet.2022.109246 .

Thomas, A., & Manuel, E. M. (2016). Embedment of Montgomery Algorithm on Elliptic Curve Cryptography over RSA Public Key Cryptography. *Procedia Technology, 24*, 911–917. https://doi.org/10.1016/j.protcy.2016.05.179

Tosun, O. K. (2021). Cyber-Attacks and Stock Market Activity. *International Review of Financial Analysis, 76*, 101795. https://doi.org/10.1016/j.irfa.2021.101795

Tsochev, G., Trifonov, R., Nakov, O., Manolov, S., & Pavlova, G. (2020). *Cyber Security: Threats and Challenges*. [online] IEEE Xplore. https://doi.org/10.1109/ICAI50593.2020.9311369

Leveraging AI and Blockchain for Space Mining

David V. Kilpin and Hamid Jahankhani

Abstract Space mining is a pioneering concept that involves extracting valuable resources from celestial bodies, such as asteroids, moons, and planets. This innovative approach holds immense potential for addressing earth resource shortages and supporting future space exploration endeavours. As humanity looks beyond our planet for sustainable solutions, space mining has emerged as a key frontier, offering a unique opportunity to tap into our solar system's vast wealth of resources. This chapter draws on prior research in the context of terrestrial mining to consider how the implementation of blockchain can introduce immutability to the management of data related to the ownership and productivity of mining. The chapter contends that the same principles of immutability can also be leveraged to explore new areas of space mining to enhance confidence in ownership through clear celestial demarcations and encourage investment through robust assessment of the productivity of extraction activities, thereby overcoming challenges and disputes that have historically shaped mining endeavours on earth.

Keywords Space mining · AI · Blockchain · Celestial exploration · Resource mapping · Predictive modelling

1 Introduction

The key components of space mining include identifying suitable targets for extraction, developing advanced technologies for mining and processing, establishing efficient transportation methods, and addressing legal and ethical issues. The primary targets for space mining are asteroids, rich in precious metals and minerals, and moons and planets that may contain valuable resources such as water and ice. Space

D. V. Kilpin · H. Jahankhani (✉)
Northumbria University, London, UK
e-mail: Hamid.jahankhani@northumbria.ac.uk

D. V. Kilpin
e-mail: david.kilpin@northumbria.ac.uk

H. Jahankhani et al. (eds.), *Space Governance*, Space Law and Policy,
https://doi.org/10.1007/978-3-031-62228-1_6

mining involves the extraction of valuable minerals, metals, and other resources from celestial bodies, such as asteroids, moons, and planets. This emerging field is driven by the growing demand for rare and essential materials on Earth and the potential to support future space exploration and colonisation efforts. Space mining opens new possibilities for resource exploitation beyond our planet and has garnered significant interest from government space agencies and private companies.

Asteroids are small celestial bodies, often rich in metals, such as gold, nickel, iron, and platinum, and are considered prime targets for space mining. They can vary in size and composition and offer diverse resources. The moons of certain planets, such as the Moon or Mars' moons Phobos and Deimos, may contain valuable resources, such as water ice, rare metals, and minerals. Asteroids contain valuable metals, such as platinum, gold, and palladium, essential for Earth's electronics, medical devices, and other high-tech industries.

Blockchain technology provides a decentralised and transparent ledger that can be used to record and verify transactions related to space mining. Smart contracts, self-executing contracts with terms directly written into the code, can automate and enforce agreements between different parties, ensure transparency, and reduce the risk of disputes. Blockchain enhances the traceability of resources by creating an immutable record for each transaction in the supply chain. This is crucial for space mining as it enables stakeholders to track the origin, processing, and transportation of resources, ensuring the integrity and authenticity of the supply chain.

The decentralised and tamper-resistant nature of blockchain enhances the security and integrity of data collected during space mining operations. This is particularly important when dealing with sensitive information related to resource composition, mining strategies, and proprietary technologies. Blockchain can facilitate decentralised governance models for space-mining operations. Blockchain-based voting systems and consensus mechanisms can democratise decision-making processes, allowing stakeholders to make crucial decisions regarding mining strategies, resource allocation, and environmental policies.

AI systems can analyse large datasets in real time and provide valuable insights into mining operations. This includes optimising the extraction processes, predicting market demands, and adapting strategies based on changing conditions. This real-time analytics capability enhances the overall efficiency of space mining. Space mining operations require careful management of energy resources. AI algorithms can optimise energy consumption by adapting to variable conditions, managing power distribution, and ensuring that energy is efficiently used in resource extraction and processing.

By combining the capabilities of AI and blockchain technologies, space-mining operations can achieve greater efficiency, transparency, and governance. These technologies contribute to the sustainable development of space resources, minimise operational risks, and create a more collaborative and secure ecosystem for the emerging space industry.

In this article, I argue that drawing parallels between space mining and the terrestrial mining sector can offer valuable insights into the potential benefits and challenges associated with adopting advanced technologies, specifically artificial intelligence (AI) and blockchains, in terrestrial mining operations. While the two contexts may differ in scale and scope, these technologies can help address and promote efficiency, transparency, and improved governance in both settings.

Identifying and extracting resources from celestial bodies requires precise resource management to optimise the efficiency of space mining operations. AI can be used to analyse data from space probes and telescopes, aiding in identifying promising targets and optimising mining processes. Similarly, in the terrestrial mining sector, AI can be employed for geological exploration, resource mapping, and predictive modelling. This can enhance the accuracy of mineral resource assessments, leading to more informed decision-making regarding mining sites and resource extraction. Autonomous robots equipped with AI can efficiently perform mining tasks in space, thereby reducing the need for direct human involvement. This enhances operational efficiency, especially in challenging and remote space environments.

AI-powered technologies, such as autonomous vehicles, drones, and remote sensing equipment, can be deployed in terrestrial mining operations to improve efficiency, reduce operational costs, and enhance worker safety. Blockchain technology can establish transparent and traceable supply chains for extra-terrestrial resources. This ensures that the origins and ownership of mined materials are verifiable, thus addressing concerns related to the legitimacy and ethical sourcing of space resources. In a terrestrial context, blockchain can enhance transparency by creating an immutable record of transactions, ensuring the authenticity of mineral supply chains. This is particularly relevant for addressing conflict minerals and promoting responsible mining practices.

Establishing clear governance frameworks for space-mining operations is crucial. Blockchain can contribute to secure and transparent record-keeping by providing an auditable trail of activities to ensure compliance with international regulations. In terrestrial mining, blockchains can enhance governance by creating a tamper-resistant ledger for contracts, agreements, and regulatory compliance. This can reduce corruption risk and promote accountability in the mining industry.

AI and blockchain technology can be used to monitor the environmental impacts of space mining activities. This includes tracking the changes in celestial bodies and ensuring responsible extraction practices. Similar technologies can be applied to monitor the environmental impact of mining operations, support sustainable practices, and comply with environmental regulations. By drawing parallels between space mining and the terrestrial mining sector, there is an opportunity to leverage the lessons learned and innovations developed in space exploration to address the challenges on Earth. AI and blockchain technologies can contribute to more sustainable, transparent, and efficient mining practices, benefitting terrestrial and extra-terrestrial resource extraction.

2 Background and Literature Review

A blockchain is a growing list of records or blocks linked using cryptography. A cryptographic hash is an algorithm that maps data of arbitrary size to a bit array of fixed size. Each block contains the previous block's hash or a form of transactional data. The purpose of obtaining information about the previous block is to reinforce each additional block with the preceding block. Blockchains are typically managed in a peer-to-peer network for use in a distributed ledger, where nodes collectively adhere to a protocol to communicate and validate new blocks. The fundamental advantage of blockchain, which may be useful for its relevance to land title management, is that it is secure by design, as data in any block cannot be retroactively altered without altering subsequent blocks (Iansiti & Lakhani, 2017). The chain that joins the two blocks provides an iterative process and ensures the integrity of the other blocks. Data integrity is assured after the block is digitally signed (Knirsch et al., 2019). This means that blockchain provides a foolproof means of storing and retrieving data, which is impossible in many developing countries' manual storage of land titles.

One question arises regarding the digital identity of those who hold land. In the digital space, identities are constantly negotiated rather than fixed. Social scientists have pointed out that digital identity shapes perceptions of physical identity. Embodied forms of digital identity have led some scholars to suggest that cyberspace liberates people from their bodies and blurs the boundaries between humans and technology (Marwick, 2013). It has also been argued that digital identity frees individuals from distinctions of race, class, gender, etc. In this context, it would make sense to consider the notion of Self-Sovereign Identity (SSI) as an approach that gives individuals autonomous control over their digital identity (Ferdous et al., 2019). According to SSI, there are entities whose identities are generated which correspond to the former. Simultaneously, identities form the basis of the attributes or identifiers. Thus, we can observe how SSI ensures anonymity while ensuring trust is established. This is because one party often provides credentials to another party and the recipient they can trust, and the trust of the verifier is then transferred to the credential holder (Sedlmeir et al., 2021).

Decentralised identifiers (DIDs) are often used to generate SSIs. These DIDs enable a verifiable decentralised identity. Traditionally, cadastres have dealt with properties in their capacity to provide a comprehensive recording of real estate. Cadastral surveys document property boundaries. A cadastral parcel is a land with unique property rights (Dale & McLaughlin, 2011). The paperwork they are generally accompanied by makes cadastral maps particularly open to politicking or manipulation compared to the seamless authentication process offered by blockchains. Specific provisions related to jurisdictions shape the management and authentication of documents. These jurisdictions shape the stakeholders required to sign and authorise documents. While, in principle, the exact mechanism can occur through blockchain via decentralised identities, traditional cadastral regulations often make it a tiresome process requiring multiple signatures, also creating storage problems.

As has been shown in the case of post-colonial societies, documents related to titles are accompanied by paperwork that can be manipulated to make claims to land, such that the government's claim may vary from the individual (Hull, 2012). There is also an element of surveying, shaped by legislation, determining the form of these competing claims. While there have been calls to digitise and introduce blockchain to reduce corruption, Konashevych (2020) has made the case that individuals should possess the right to choose whether they prefer traditional registries of blockchains, arguing from a shift from cadastre bodies as having a monopoly on the market. Scholars have referred to using blockchain in land regulation as real estate tokenisation (Konashevych, 2020). Moreover, it has been argued that it is important to have a cross-blockchain protocol designed to support free choice and transferability because blockchain offers a disadvantage related to intolerance to retroactive transactions, as mentioned above.

Aditya et al. (2021) made a case for participatory land registration, which offsets the perception that government surveying may lack accuracy, thus leading to competing claims. Asiama et al. (2003) also considered land consolidation through participatory surveying to reduce the risk of land segmentation, which may create food security problems, showing that land management is also tied to broader social issues. Ameyaw and de Vries provided a Ghanaian perspective and showed that blockchain technology can enhance the transparency of land administrative functions, especially in contexts with excessive red tape. Other scholars have joined the debate on the intersection between land management and blockchain, such as Bennett et al. (2021), who showed how blockchain technologies can significantly impact smart contracts.

Using case studies from multiple contexts, they examined hybrid approaches to introducing smart contracts. The results show that the hybrid approach can enable adherence to land administration business requirements and generate greater trust. In many contexts, there is suspicion of centralised government land management systems, given longstanding colonial histories that still impact national land management systems in many African countries. In this context, many scholars have referred to blockchain as an anti-corruption tool, with Aarvik (2020) suggesting that the introduction of blockchain directly affects many areas of governance and trust. Aarvik (2020) suggested that blockchain is likely to work more effectively in areas where trust in codes is likely greater than trust in persons, yet has made the case that digital literacy is necessary before implementing blockchain.

3 Overview of Existing Literature on Space Mining, Focusing on Technological, Legal, and Ethical Considerations

The literature on space mining underscores the formidable technological challenges inherent to extracting resources from celestial bodies. Researchers have developed advanced robotics and autonomous systems tailored to harsh space conditions. Autonomous mining probes equipped with artificial intelligence (AI) have emerged as key components capable of prospecting, extracting, and processing resources on distant asteroids and moons. The need for innovative mining techniques and machinery has led to discussions on adapting terrestrial mining technologies and developing novel methods suitable for low-gravity environments. Exploring space-based 3D printing and in situ resource utilisation (ISRU) technologies is also prominent, as these innovations could facilitate on-site construction and utilisation of extracted materials, reducing the reliance on Earth for supplies.

A significant portion of the literature on space mining is dedicated to the legal framework of extra-terrestrial resource extraction. Scholars have highlighted the existing international space law, such as the Outer Space Treaty, and discussed its applicability to property rights and resource ownership in space. Debates regarding the need for updates to international laws to accommodate the unique challenges posed by space mining are common. Attention is paid to jurisdiction, liability, and establishing a clear legal regime that governs the rights and responsibilities of entities engaged in space mining. The literature emphasises the necessity of international collaboration to develop a comprehensive legal framework that fosters responsible and sustainable space-mining practices.

Ethical considerations in the space mining literature revolve around the responsible and sustainable exploitation of extra-terrestrial resources. Scholars have explored the potential environmental impacts of space mining activities, emphasising the importance of minimising harm to celestial bodies and maintaining the integrity of space ecosystems. Discussions also delve into the ethical sourcing of materials, ensuring that space mining operations adhere to standards akin to those increasingly demanded by terrestrial industries. Additionally, there is a focus on addressing potential conflicts arising from resource competition and the distribution of benefits, urging the establishment of ethical guidelines to promote fair and equitable practices for space resource extraction.

The literature on space mining offers a comprehensive exploration of the challenges and opportunities in this evolving field. Abbud-Madrid et al. (2016) presented seminal work exploring the technological, economic, legal, and ethical dimensions of space mining. This emphasises the imperative for innovation in mining technologies, the necessity of international collaboration, and the establishment of regulatory frameworks to govern extra-terrestrial resource extraction. A subsequent contribution by Abbud-Madrid (2021) focuses on the economic aspects of space mining. This study assesses the evolving space economy and highlights the potential economic benefits of space mining and its pivotal role in supporting space exploration and development. Badescu and Zacny (2018) comprehensively explore the scientific, technical, and practical aspects of space mining. It delves into the types of celestial bodies targeted for resource extraction, mining technologies, and the potential impact of space resources on future space exploration and industries. Xie et al. (2022) contribute a survey paper that offers a technical perspective on asteroid mining. It delves into the challenges associated with extracting resources from asteroids, including prospecting, mining methods, and using mined materials.

Turning attention to legal and ethical dimensions, Goswami et al. (2020) address the current state of international space law and its implications for space mining. This work underscores the need for legal frameworks governing property rights, liabilities, and regulatory considerations in space resource extraction. Sonter (1998) first evaluated the technical feasibility and economic viability of space mining. This study assesses various factors, including mining technologies, resource utilisation, and economic models, providing insights into space resource extraction's potential challenges and benefits. These studies contribute to a nuanced understanding of space mining, covering technological intricacies, economic implications, legal considerations, and ethical dimensions. As the field evolves, staying abreast of recent articles, conference proceedings, and dedicated journals remains crucial to comprehending the latest space exploration and mining developments.

In the quest for space mining advancement, researchers have explored cutting-edge technologies poised to redefine the industry. Sercel et al. (2018) highlighted the potential for in-space resource utilisation. This paper envisions a future in which robotics and mining technologies evolve in tandem, with private industries playing a pivotal role in driving innovation. Many researchers have argued for the introduction of deep reinforcement learning for autonomous asteroid prospecting, showcasing the role of artificial intelligence in enhancing the autonomy of space-mining missions (Paek et al., 2019).

As space mining activities gain momentum, legal frameworks have become central to governing extra-terrestrial resource extraction. Studies have also considered the implication of policies such as the "Commercial Space Launch Competitiveness Act (CSLCA)", but there generally appears to be a lacunae around legal frameworks for space mining (Christensen et al., 2019). This work underscores the importance of evolving legal landscapes and the need for legislation to catch up with the burgeoning space-mining industry. Additionally, Lee's exploration of property rights in outer space delves into the complex legal considerations surrounding property rights in the context of mining activities in outer space.

Ethical considerations form a critical component of the discourse on space mining. Scholars have also explored the ethical landscape of space mining by examining environmental impact, corporate responsibility, and potential societal implications. Acknowledging the necessity for responsible resource utilisation, Hornsey et al. (2022) addressed the ethical dimensions of space activities, emphasising the importance of sustainable practices.

4 The Current State of AI and Blockchain Technologies in Terrestrial Mining Operations

Existing research on mining in Africa has shown that private companies have far greater power over local governments in shaping mining regulations, which has resulted in the marginalisation of local owners. Owing to the incapacity to monitor and implement regulations, the result may be bypassing government regulations by private entities, who, in the past, have also co-opted armed groups to protect their installations. Land use conflicts between large-scale miners and community members have been frequent worldwide. Consequently, it has become increasingly difficult for miners who demand a significant area to co-exist with community members, especially as they rely on these lands for their livelihoods (Hilson, 2002). It has also been noted that government involvement in dispute resolution is often miniscule. Thus, in the past, it has been recommended that if governments play a central role in arbitrating conflicts, they can allow for improved dispute resolution. As Castro and Neilson (2001) have shown, natural resource conflicts are often "severe and debilitating, resulting in violence, resource degradation, the undermining of livelihoods, and the uprooting of communities".

Often, when utilising natural resources such as mines, parties try to maximise their interests and, in the process, end up defeating the interests of others (Odhiambo, 2000). Mining often has wide-ranging impacts on communities. For example, for an indigenous community which has inhabited a tract of land for several decades, a 20- to 30-year-old mining project is seen as "a disruption in the flow of time, much like the effects of an earthquake or major flood" except that the effect can be much slower and enduring (Asp, 2000, 10). Yet, among developing countries in Africa, such as Zambia, mining contributes 90% of the country's export earnings. It employs 15% of the country's workforce, meaning that it is often in the government's interest to incentivise mining companies over the interests of the communities (Mbendi, 1999).

In many countries, mining efforts have also meant that all communities have effectively come to live under a country's mining project, as in the case of the Northern Brazilian state of Para, where 400,000 individuals seek to tap into the residents' resources, 100,000 of whom live directly inside the gold mine (Lemieux, 1997). Another way mining affects a local community is when demographic shifts caused by mining increase the cost of living, resulting in indigenous communities rapidly seeking to derive livelihoods in other areas (Dorian & Humphreys, 2000). There is a possibility of conflict between landowners and owners of mining companies, as well as between miners and herders, as mining introduces new ideas of capital ownership with which local communities are unfamiliar.

Countries like Ghana have a history of conflict between small miners and large-scale mine management. Individual mine lands are often located illegally within the land occupied by large-scale mining companies, often the product of familial and cultural relations with the land. Ayapong has shown that chasing small miners off their exploration areas has become increasingly difficult. When the army or the police are called upon, violence is typically the result. For instance, in Fiji, indigenous people believe they own everything above or below the ground (Mcleod, 2000), whereas in New Guinea, 90% of the land is owned by residents (MiningWatch, 2000).

The various types of mining property include mineral property, unified estate, severed or split estate and fractional ownership of minerals. Importantly, in a unified estate, the surface and mining rights are not severed, which means that there is an assumption that owning the land also translates into owning the mine. In severed or split estates, mineral and surface ownership is severed. Essentially, this means that while companies may own the mining rights, the land may be owned by private individuals. Lastly, fractional ownership refers to instances where the ownership of the mining property is split between several individuals. The complexity raised by these categories is such that they show that ownership is not consolidated in a single individual, resulting in conflict between different owners and private mining companies.

This leads to the question of how mineral rights are severed from surface estates. There is often a severance when the surface property or the mineral is sold, with the other being retained. Yet, it is often common for the surface to be sold and the minerals to be retained (Kent, 2019). When there is a split estate, the mineral rights owner often has the right to develop the mineral resources. Ryder and Hall (2017) demonstrated that due to technological advancements, extraction processes have now been expanded to areas without oil or gas development. Another reason why mineral rights are such a contested issue is because of the so-called rule of capture, according to which minerals which have the capability of migration beneath the Earth's surface can be extracted even if the source was someone else's property. Despite this, the state often mediates conflicts between mineral owners.

Given the concept of split ownership, this situation forces us to consider the possibilities and challenges that exist with the implementation of blockchain. To answer this question, it would first be useful to consider the challenges of digitisation alongside the opportunities it creates for actors. Partial digitisation has been more time-consuming as it appears technically complicated and requires process-specific training for technical and non-technical employees (Deshpande, 2019). Information systems that enhance learning and knowledge management have mixed effects, often depending on organisational and other specifics (Huber, 2001). For example, access to knowledge management systems strengthens the association between innovation culture and innovation outcomes but diminishes the association between an autonomous culture and innovation outcomes (Durcikova et al., 2011). Relatedly, using IS to enhance knowledge capabilities may enable continuous innovation (Joshi et al., 2010).

We may take the lead from some of the literature on using blockchain to assess geological changes. This literature may be useful as it has shown how blockchain may be used to assess changes in land characteristics in exploration projects. As mentioned above, the cadastre is often accompanied by extensive paperwork and lacks the flexibility to accommodate real-time changes, as a land survey is important. Abramov et al. (2021) have suggested using innovative technologies within geoinformation management, arguing for geoinformation support in the context of climate change. Insights can also be gleaned from the oil and gas industry about blockchain utilisation and the increased efficiency it can bring. Just as the literature has shown the immense potential of blockchain to revolutionise land management, there has been relative silence concerning the problems of mining land management. For example, split ownership has often resulted in the rights of private mining companies superseding that of landowners. This problem may further be exacerbated when property transactions are stored as printed copies as legal proofs for a long period. This does not account for the changes in agreement between the contracting parties, where there is immense potential to introduce blockchain, where each sale or transaction is linked to a block tied to a previous digital signature. This may have important implications for local communities in developing countries, where land is held by extended families who rely on it for their income.

Given the digitisation problems without adequate training in much of the developing world, Alam et al. (2020) make a case for incremental digitisation or the implementation of blockchain in land title management in Bangladesh. They propose a blockchain-based solution that offers data synchronisation and transparency, ease of access, and immutable records management as a faster and cheaper solution. They state, "Considering the technological knowledge and capacity of the people and the government, we introduced a phase-by-phase Blockchain adoption model that starts with a public Blockchain ledger and later gradually incorporates two levels of Hybrid Blockchain" (Alam et al., 2020, 1319).

The resulting decentralisation of control, in combination with the immutable representation of the transfer of possession, provides a unique opportunity to build collaborative, multi-sided, 'trustless systems' (Glaser, 2017). Ølnes et al. (2017) enumerate the benefits and promises of blockchain technology in government as a means of information sharing. In general, blockchain technology allows for client-defined chains of entries, client-side validation of entries, a distributed consensus algorithm for recording entries, and a mechanism for ensuring security. The data is secured by publishing it in an encrypted form in the immutable, distributed ledger (Nakamoto, 2008). The storage of critical information in the decentralised data layer protects the system from malfunction, obsolescence, and deliberate or involuntary record changes. All critical data records are secured by a "proof of existence" in the blockchain (Lemieux, 1997).

Every essential step in the registration process that qualifies as a high "standard of care" can be published to the blockchain. This provides data security and long-term stability by making the entire data set immutable via notarisation. In addition, publishing data during each step of an instrument's recording process provides irrefutable "proof of process", protecting the submitted records and establishing accountability. It is important to note that blockchain technology does not help address the accuracy of the land titles (Ølnes et al., 2017). According to Mougayar (2016), 80% of the work in adopting blockchain technology is associated with changing business processes, and only 20% is related to technology implementation. Most of this transformation happens in the background because blockchain is an invisible enabler that does not affect externally visible parts of a business or organisational entity. Just as there is a potential to leverage blockchains for terrestrial mining, a sound understanding of the uses of blockchains can also be incredibly useful in any future endeavours related to space mining.

5 Gaps in the Literature, Particularly Regarding the Application of These Technologies in Space Mining

While the existing literature on space mining provides valuable insights, notable gaps highlight areas for further research and exploration. These gaps centre around the application of technologies, specifically artificial intelligence (AI) and blockchain, in the context of space mining. While there is an acknowledgement of the role of AI in autonomous space mining operations, there is a gap in detailed studies on integrating AI with robotics. Future research could delve into optimizing the synergy between AI algorithms and robotic systems, addressing challenges related to real-time decision-making, adaptability to varying conditions, and enhancing the overall efficiency of mining processes. The literature provides a broad overview of potential mining technologies, but there is a need for more in-depth studies on adapting existing terrestrial mining technologies for extra-terrestrial environments. Research could focus on modifying and optimizing equipment for low-gravity conditions, extreme temperatures, and the unique challenges presented by celestial bodies. The current literature touches upon AI applications in resource prospecting, but there is a gap in detailed studies that explore the full potential of AI in this phase of space mining. Research could investigate the development of AI algorithms for analysing remote sensing data, identifying promising mining sites, and optimizing resource prospecting strategies. While there are discussions on the use of blockchain for transparency in space mining supply chains, there is a gap in exploring the full extent of blockchain's capabilities in ensuring traceability and governance. Future research could delve into developing blockchain-based systems for immutable record-keeping, smart contracts, and decentralized governance structures to address legal and ethical considerations. The literature mentions environmental impact considerations, but comprehensive studies have a gap in assessing the long-term environmental consequences of space mining activities. Future research could focus on developing AI models for monitoring and predicting the environmental impact of mining operations in space, ensuring sustainability and responsible resource utilization. While there is recognition of the role of AI in autonomous operations, there is a gap in understanding the dynamics of human–machine collaboration in space mining. Research could explore how AI systems can complement human decision-making, particularly when human intervention is necessary, and how to optimize collaboration for enhanced efficiency and safety. The legal literature acknowledges the need for regulatory frameworks. Still, it lacks an in-depth exploration of the specific regulatory challenges posed by emerging technologies like AI and blockchain in space mining. Future research could focus on proposing regulatory guidelines that address the unique aspects of these technologies, ensuring responsible and ethical space mining practices. Addressing these gaps in the literature will contribute to a more comprehensive understanding of the application of AI and blockchain technologies in space mining, fostering innovation and sustainable development in this evolving field.

6 The Methodological Approach for Analysing the Potential Application of AI and Blockchain in Space Mining

Analysing the potential application of artificial intelligence (AI) and blockchain in space mining requires a comprehensive methodological approach. This article employs a comparative methodology to consider the usefulness of blockchain implementation in the mining sector in expanding the understanding of space mining. Subsequently, a technology assessment will be conducted, evaluating the capabilities and limitations of AI and blockchain in space mining. A technical feasibility study follows, examining how these technologies can be adapted to the unique conditions of space environments. Simultaneously, an economic viability analysis should be undertaken to gauge the financial implications of integrating AI and blockchain, considering cost savings, operational efficiency, and potential returns on investment. Legal and regulatory considerations become pivotal, examining the existing space law framework and proposing or assessing regulatory models for these emerging technologies. Ethical impact assessments should address environmental and societal concerns, ensuring responsible and sustainable resource extraction practices. Including case studies, prototyping, stakeholder engagement, risk analysis, and iterative refinement completes this methodological framework, providing a holistic understanding of the potential applications of AI and blockchain in the dynamic field of space mining.

7 Description of the Comparative Analysis Framework Used to Integrate Findings From the Terrestrial Mining Sector

The comparative analysis framework employed to integrate the terrestrial mining sector findings into the context of space mining involves a multi-faceted examination. Beginning with assessing mining technologies, the framework scrutinizes the adaptability of terrestrial mining methods to space environments and identifies technological parallels and disparities. The focus extends to resource management, evaluating how artificial intelligence (AI) can optimize terrestrials and extra-terrestrial resource extraction. Operational efficiency and automation are compared, investigating how AI-driven automation enhances productivity and safety in both sectors. The analysis extends to transparency and traceability, where the existing mechanisms in the terrestrial mining sector are contrasted with the proposed blockchain applications in space mining for responsible material sourcing. Governance structures, legal frameworks, and environmental impact considerations are scrutinized, drawing comparisons between terrestrial mining's regulatory landscape and the emerging legal considerations in space mining. The framework also encompasses economic opportunities, job creation, and technological innovation, exploring synergies and collaborative potentials between terrestrial and space mining endeavours. This comprehensive approach ensures a nuanced understanding of how lessons from the terrestrial mining sector can inform and contribute to the nascent field of space mining technologies.

8 Potential Application of Blockchain for Secure and Transparent Record-Keeping in Space Mining Cadastres

The application of blockchain technology in space mining cadastres holds immense potential for revolutionizing the management of extra-terrestrial resources. By leveraging its decentralized and tamper-resistant architecture, blockchain ensures the creation of an immutable record-keeping system, establishing trust and authenticity in recording ownership, claims, and transactions within the space mining cadastre. Smart contracts, embedded within the blockchain, automate various processes such as lease agreements and permit renewals, streamlining administrative tasks and reducing the risk of errors. The transparency inherent in blockchain facilitates regulatory compliance, allowing real-time monitoring of space mining activities and reducing the likelihood of fraudulent practices. Additionally, blockchain enhances data storage security, safeguarding critical information related to geological surveys and ownership records. This technology's ability to trace the origin and movement of extracted materials promotes responsible sourcing and ethical practices in space mining, contributing to a sustainable and transparent framework for the exploration and utilization of celestial resources.

9 Analysis of AI's Impact on Enhancing Operational Efficiency in Mining, With Insights From Terrestrial Mining

When participants were asked whether they thought digital technologies like blockchain would automatically change behaviours and increase compliance with rules and regulations of the mining sector, many of them shared the importance of maintaining independence between technological innovation and digital literacy. Most did not think that innovations would solve the problem alone. Miners and government officials often lamented the low levels of digital literacy. They shared that it was essential to educate miners and government officials about basic digital technologies, cadastre, and blockchain without its implementation to automatically result in behaviour change and improved compliance. This was contrary to much of the existing discussions on implementing digital technologies and their impact on behaviour, which assumes that people will automatically change behaviours. Administrative problems are likely compounded without adequate training, resulting in people finding new ways to dodge digital technologies and engage in corrupt practices. Moreover, digitisation without proper training will also result in managerial mismanagement, where existing asymmetries and degrees of uptake of digitisation between departments are likely to become compounded. Innovation laboratories, incubators, and new digital products are examples of digital experiments used in the early stages of digital transformation that achieved impressive outcomes (Desmet et al., 2015). However, digitisation profoundly impacts every aspect of modern business. Each digital endeavour necessitates a previously unheard-of collaboration across people, systems, and tools. Organisations are under tremendous pressure to change as a result of digitisation. As a result, to become a digital organisation, substantial changes must be made to how businesses function and conduct (Dorner & Meffert, 2015). Expanding these competencies while incorporating them into the core business is an important question. None of the organisational changes these technologies bring occur without managerial mindsets, organisational behaviour, or operating cultures. Only organisational transformation can result in better performance from digital technologies. Digital transformation is fundamentally about change, and like any organisational change program, it needs to be handled carefully (Wade, 2015). Digital transformation is also a portfolio of efforts that work together to scale change rather than a single endeavour. A "let a hundred flowers bloom" strategy, in which each initiative operates separately and with an experimental mindset, can produce compelling and intriguing outcomes in all areas of ideation and innovation. Still, it is not a method for scaling a digital transformation throughout an organisation. Management's coordination of activities that run concurrently ultimately determines success (Dorner & Meffert, 2015). New organisational competencies and skill sets are required to enhance the adoption of digitalisation throughout a company. Managing the shift to a digitally driven business model is essential for survival and outperforming rivals. The effects of digitisation on companies and those who work in them must be understood by the program leaders for digital transformation. Launching digital transformation programs to apply these digital solutions in enterprises has become increasingly prevalent due to the extensive availability of digital technologies and apps. Digital transformation refers to "the increasing adoption of digital tools and technologies by an organisation to fundamentally alter its internal and external processes and functions" (Bonnet & Nandan, 2011, p. 2). It is referred to as digital transformation. Wade (2015, p. 2) notes that in the organisational context, digital transformation typically has a specific goal or 'Value Proposition' that is added and defines it as "organisational change through the use of digital technologies to improve performance materially" (see also Gimpel & Roglinger, 2015). Digital transformations are built on the foundation of digital technology to achieve business benefits. Digital transformations can be distinguished from alternative company transformations, such as turnarounds or reorganisations. However, it is currently standard practice for digital transformation efforts to place an undue emphasis on technology (Bonnet & Nandan, 2011). Such a technology-focused strategy frequently falls short of challenging the foundation of a business model. The approach's focus on automating current business processes rather than trying to re-engineer them to support new business models entirely is one reason for this. Furthermore, technology-focused transitions frequently severe IT ties to business operations, which increases people's aversion to change (Bonnet & Nandan, 2011). Businesses are putting money into digitisation without trying to force necessary adjustments. They fail to recognise how digitisation would affect organisational structures, human dynamics, organisational cultures, and technology since "it is not technology that is an obstacle to digital transformation, it is people' (quoted in Bonnet & Nandan, 2011, p. 8). New collaboration methods with clients and partners, as well as within businesses, are made possible through digital technology and applications.

Medical experts have considered using blockchain technology to improve global health by providing universal access to funding mechanisms and smart contracts, ensuring patient privacy, improving referrals to specialists in the region, and ensuring new standards for payments and reimbursements in the Caribbean to combat COVID-19 (Resiere et al., 2020) suggest that supply chain applications are also used to connect sellers to buyers of low income. Online cocoa trading platforms for private enterprises include the UTZ Good Inside Portal, a voluntary certification scheme. Ghana's cocoa supply chain could benefit from large-scale adoption, which can be evaluated as a part of COVID-19 and post-COVID-19 management. Several factors influence whether and how the benefits of digitising land records have been realised (Moreri et al. 2018). Land records cannot be digitised in a supportive environment. Studies assessing the impact of digitising land records in GS have produced mixed results. Various benefits of the EDMS implemented in Lagos, Nigeria, were identified. It is easier to find and retrieve documents by centrally storing land-related files. Administrative efficiency increased due to the 40 "Training is very important, even in the manual system, it is very important. So you have people coming and starting employment. They also need all the job training. And they will all start every day and then work for many years. So yeah, training is very important, especially at the moment when you newly hire someone (Participant 10)." reduced waiting time for land-related information. Public confidence has generally risen (Thontteh & Omirin, 2015). Another example is Bhoomi, a government-funded initiative in India's Karnataka State (Benjamin et al., 2007). Twenty million land records owned by 6.7 million farmers were entered into a computerised database. Due to the socio-legal processes underpinning land claims in the pre-Bhoomi era, many different land ownership types existed. With Bhoomi, land ownership forms were standardised. A primary cause of land-related corruption is ambiguity in land documents, as in the pre-Bhoomi era, which was controlled mainly by local street bureaucrats (Benjamin & Raman, 2011). Digitalisation helped overcome some of these problems. Bhoomi and the EDMS have limitations and drawbacks. For example, the EDMS did not improve the outcome of boundary disputes. Additionally, the EDMS system did not increase the application processing or revenue (ibid.). Despite its benefits to farmers, the digitisation of land records is not necessarily associated with improved welfare, according to an analysis of the Bhoomi system. The cropping patterns for the previous 12 seasons were also included in the records. The digitisation of land cadastral records was arguably the largest ever undertaken in a developing country. Benjamin et al. (2007) attribute the Bhoomi system's rise to corruption, bribes, and the long time it took to complete transactions (Benjamin et al., 2007). Digitisation was achieved by centralising land ownership records, so management shifted from villages to district-level taluk offices. Before Bhoomi, it would typically take two to three days to receive a copy of a Record of Rights, Tenancy, and Crop Information (RTC) or mutation–the transfer of rights from one owner to another. These documents were efficiently made available by village accountants (VAs). As a result of Bhoomi's implementation, farmers had to depend on agents to perform their tasks, and mutations could take up to four months. Farmers often found no electricity or computers when they visited the Taluk office, which caused them to miss several farming days. Land records were centralised and computerised to benefit large- and middle-level farmers. They allegedly accessed the government benefits provided to farmers, such as subsidies for seeds, fertilisers, and pesticides, using the survey numbers of small farmers. In addition to acquiring land in prime locations from small farmers at a low price, Bhoomi assisted powerful economic groups in acquiring land at a low cost. The farmers were often given these lands under various 'inam' schemes, which restricted their use of selling land. Several small landowners in an area were approached by economically and politically connected real estate agents who negotiated with them and converted the land for commercial use. They obtained a No Objection Certificate (NOC) from the village panchayat and used the Bhoomi system to issue titles to their clients. Developers had connections to government officials and politicians. Efforts to influence the process or make decisions were impossible in such cases because powerful and influential interests had a hand in the process. This explains why any implementation of blockchains requires a robust understanding of power structures and the interests of relevant stakeholders, specifically whether they support or oppose the digitization of the mining sector.

10 Exploration of AI's Potential in Optimizing Mineral Identification and Extraction Processes in Space Mining

The discussions around digitization and blockchain in the mining sector have illuminated critical considerations, particularly regarding the need for digital literacy and the potential impact on behavioural change. Drawing parallels with these insights, the application of AI in mineral detection and extraction processes in space mining introduces a new dimension to the discourse. The consensus among participants on the importance of education and literacy in embracing technological innovations is equally relevant in the context of space exploration. The integration of AI in mineral detection on celestial bodies necessitates advanced technological infrastructure and a workforce equipped with the necessary digital skills. Space agencies and mining entities engaging in extra-terrestrial resource extraction must prioritize comprehensive training programs to bridge the gap between existing skill sets and the demands of AI-driven exploration.

Furthermore, the parallels between the challenges faced in the terrestrial mining sector, such as managerial mismanagement and resistance to change, find resonance in the space exploration domain. Space agencies and industry players must carefully manage the organizational change that accompanies the integration of AI technologies, ensuring that leadership embraces a digital mindset and fosters a culture of innovation. The lessons learned from the digitization experiences in Earth's mining sector underscore the critical importance of addressing educational needs, promoting a forward-looking organizational culture, and navigating the complexities of technological integration, all equally pertinent in the unfolding era of space mining and exploration.

11 Discussion on the Challenges and Opportunities of International Regulatory Compliance for Space Mining, Informed by the Adoption of AI and Blockchain in Zimbabwe

The participants also shared the need for legal changes accompanying digitisation. Many shared that legal changes had already been made, but in the absence of enforcement, legal changes were likely to be dead letters. However, "Before 2004… that's when people had the idea of a computerised mining ministry. It was first started, and by the end of 2022, some people in the ministry went to Canada to see how this could be done. Of course, we live in a global village. Each of our stakeholders, our clients from the region and overseas, should be involved in how we implement these technologies (Participant 29)." many participants shared that digitisation could also result in greater compliance, thus making the law more enforceable. This is because if blockchain could detect infractions efficiently, people would be more likely to engage in ethical and legal practices and avoid corruption or practices, such as double pegging. The use of new technologies in government and their ability to revolutionise them have drawn much attention. Meijer (2006) describes public-sector digitisation processes as using ICTs by governmental bodies to carry out their operational and administrative procedures. According to Dobrolyubova et al. (2019), these processes are frequently considered facilitators or drivers of transformation. Attempts to digitise public services have had an impact on a variety of stakeholders, often from many spheres of government and society. In terms of digitalisation, African nations generally lag behind Western countries. Due to continuing illiteracy in the indigenous population, limited access to ICT devices and services, high usage charges, and inadequate ICT infrastructure imported ICT systems cannot be used as intended (Tchamyou et al., 2019). Nevertheless, African countries have built resilience and are catching up in the global effort to introduce new technologies in the public sector. Promising results have been found in various areas related to using ICTs, which calls for a broader and more systematic investigation of the situation on the continent (Olatokun, 2008). In this regard, ICTs can positively and significantly improve public sector management practices, as Evans (2018) outlined. In Africa, digitising government is frequently seen as a practical and effective means of improving governance. Many governments plan to employ ICTs to enhance transparency, expedite internal processes, or change how accountability and 229 openness are included in the provision of public services. Accountability enables the public to evaluate and punish representatives. Although there are various types of responsibility, they can be broadly categorised into internal and external. Public officials must report to those who oversee their work at all organisational levels to be held internally accountable. According to the assumption that governmental institutions are structured hierarchically and accountable to their political masters, this refers to hierarchical responsibility. Unlike internal accountability, external accountability entails organisations and public employees being accountable to relevant authorities outside their organisations (Siddiquee, 2005). According to Lührmann et al. (2020), in this situation, accountability for officials' acts and/or organisations' levels of performance is frequently diagonal. Civil society and media actors exert pressure on governments, mainly by providing information or making specific requests for accountability. Because workflow can be traced, ICT systems can track corrupt behaviours and monitor work processes concerning accountability. This explains why international organisations strongly support technological development and adoption in the public sector on the grounds of increased government transparency and effectiveness. A basic supposition would be that ICTs will strengthen hierarchical responsibility inside the governmental apparatus. A second supposition is that the impact of digitisation on the relationship between the state and its citizens leads to greater diagonal responsibility. Some participants, especially miners, thought it necessary to have a political will to impose laws, without which infractions and unethical practices were likely to continue. They felt that mining-related laws had to be implemented independently of digital technologies. For example, one person suggested, "There needs to be a political will to enforce things and respect the law (Participant 15)." Many participants thought there were already sufficient laws, but the major problem was their implementation. Some felt that while the ministry intended to implement new laws, it was there, but legislation to show that will be missing, as one participant stated, "Actually, the ministry has been trying to convince the government to amend the legislation for more than 20 years now. These are mines and minerals that act. Yeah. Yeah, so. The intention to amend. To a considerable extent, this shows that the Ministry believes that. However, legislation is time-consuming. It's an inconvenience for the investors and the stakeholders, and it's taking too long (Participant 8)." This showed that the Ministry needed to engage in much more serious efforts to persuade legislators, but also that any effort at digitisation was likely to fail without a concrete legislative framework which could dictate how the digital transformation process was likely to take place. Discussion on the Need for Legal Reforms Most participants suggested that if technological innovation took place without legal changes, the law itself would become an obstacle to innovation, given the discretionary ways the law works against weaker miners. When asked whether it was necessary to have legal changes before introducing a new title management system, one participant stated,' I am always a supporter of a strong legal framework and institutional framework for technology... Before making the technological changes. However, you know, you know, technology can come and go. So, you need to have a strong foundation (Participant 17)." Some participants also thought that the existing legal imagination in the mining sector was still that of the manual system, which is why it was necessary to create laws in line with the requirements of cadastre and blockchain. One participant stated that the law was based on a manual system. He said that it is vital for the law to underpin technology while also complaining about the outdated nature of laws, "The law needs to incorporate the technology also. The law must also remove some old ways of promoting the manual system (Participant, 29)."

In evaluating the legal and regulatory frameworks for digitization and blockchain in the mining sector, participants emphasized the necessity of legal changes accompanying technological advancements. They acknowledged that legal modifications were already in place, but their effective enforcement was crucial. The discussion echoed the sentiment that digitization could enhance compliance, making laws more enforceable, especially if blockchain technology efficiently detected infractions. This alignment of legal and technological developments is vital to fostering ethical practices and preventing corruption.

Drawing parallels with this perspective in space mining, the question arises: should the law precede technological change in extra-terrestrial resource exploration? The insights from participants underscore the importance of a synchronized approach, emphasizing that legal changes and technological innovations should go hand in hand. As articulated by some participants, the call for a robust legal framework aligns with the notion that technology should not outpace or undermine the legal foundation. This sentiment resonates with the view that a strong legal and institutional framework is essential before introducing technological changes, creating a foundation to support and guide technological evolution.

Furthermore, the critical review of the adequacy of current regulations in the mining sector revealed concerns about the need for political will to enforce laws effectively. This echoes the sentiment that legal changes alone are insufficient without a commitment to implementation. The discussion on the need for legal reforms in the mining sector raises parallels for space mining, emphasizing that mere technological innovation without corresponding legal adjustments could hinder progress. The assertion that existing legal imaginaries in the mining sector are rooted in the manual system calls for a proactive approach to align laws with emerging technologies like cadastre and blockchain requirements.

In summary, the discussions in the mining sector emphasize the interdependence of legal and technological advancements. This perspective provides valuable insights for space mining, suggesting that the development of a legal framework should precede or occur concurrently with technological innovations to ensure a harmonised and practical approach to extra-terrestrial resource exploration.

12 Ethical and Cultural Considerations in Implementing Technology-Driven Mining Solutions, With Lessons From Terrestrial Mining

Many of the participants shared the necessity of having an indigenous impetus for innovation, as many thought that in the absence of a local impetus, the problems were only likely to exacerbate in the absence of an inter-sector and holistic move toward digitisation, which would also entail a concomitant change in attitudes in addition to a change in operating systems. This will also entail integrating operations between different government ministries, which would be impossible if external actors initiated the discourse around innovation. There has been a long debate on whether innovations should develop indigenously or exogenously. Since the local elite are constantly present and, on the lookout, Andrea Cornwall argues that "giving voice" to the weak in public places can have disastrous results for those who speak up (2005). Additionally, I have seen how "guided discussions" facilitated by Social Innovation practitioners can subvert participatory procedures by gently guiding participants to the location set apart just for them. The procedure I referred to as a "grassroots movement" is if the neighbourhood has understood what "they want", which is compatible with what the project can provide. This allows for converting externally developed approaches or procedures into "community-based values." The potential loss of community and sense of place, which are frequently linked to economic success, is another issue that concerns Indigenous people. Academics have drawn attention to problems similar to those in developed economies like the US. Therefore, not only should we anticipate that some indigenous people may desire to maintain their old ways, but we should also be prepared for the possibility that people from advanced economies will yearn to return to something akin to what indigenous people currently experience while retaining significant modern benefits. While democratic governance and egalitarian social relations are admirable objectives for any community, it is also true that imposing these structures on target communities will not likely result in anything resembling people-centred sustainable development. Within this development area, resistance to agency presence may be attributed to the imposition of exogenous political representation. This might also indicate the underlying conflict between contemporary and traditional African ways of being and acting. Unintentionally pushing II to the periphery of the development process is an unusual tendency among development specialists. For instance, Gupta (2013) says, "Grassroots innovations emerge when existing systems and practices fail to serve people's needs." This statement implies that IIs are only considered when the "existing system" has failed if we accept that grassroots innovation is akin to an II. In other words, as long as exogenous or mainstream systems appear to address the demands of underprivileged groups, grassroots or II groups can be safely ignored. This way of thinking marginalises II, and indigenous systems are helpless when exogenous forms of involvement are driven by external change agents operating within indigenous communities. It may be argued that induced social change may not always be necessary for successful SI. That readiness to comprehend and support pre-existing institutions in their current form may be more critical. This will not help traditional communities transition quickly into modern society; on the contrary, it involves a period of growth that cannot anticipate progress toward the objectives and values of contemporary Western societies. However, a procedure started and managed by community members is more likely to result in gratifying and long-lasting outcomes for those groups. This is particularly true in cultures where men and women are segregated into different semi-autonomous communities. Almost all miners suggested that it was necessary to have the right balance between innovations led indigenously and exogenous innovations. They all thought that if innovation were to be led exogenously, it would only intensify the problems already faced in the mining sector. Out of 30, only 2 supported exogenous innovation, whereas 20 supported purely indigenous innovation. Six miners supported a mix of both indigenous and exogenous innovation. The reason for greater support for indigenous innovation was because of the miners' prior knowledge and experience that external interventions would only further complicate the systems without the necessary support from the mining community. 226 Figure 28 Support for Type of Innovation Some miners also suggested that the arrival of digital technology was not something one could demand, as digitisation was a process with inertia, given that digital technologies undergo diffusion from one country to another. Some participants suggested that miners also demanded the implement digital technologies in addition to the inertia, given the problems they faced regularly. This meant that stakeholders such as miners already had a vague idea about what digitisation would entail—they all considered it positive. Still, they often considered digitisation a process that would create structural and administrative changes. When asked what exactly they meant by implementing 37, One participant stated, "I think it's the right time [to implement digital technologies]. The technology is the driver, and the miners require that. Technology has arrived. The technology is available, but the miners also demand it (Participant 18)." Type of innovation (out of 30) Support for exogenous innovation Support for indigenous innovation Support for both digital technologies, some of them suggested the GIS. In contrast, others mentioned the cadastre and discussed the need for mobile applications to improve the allocation process. This leads us to believe that the local miners are demanding change. While they all suggest different innovations, they suppose they will slowly begin to improve administrative processing. Compared to studies where local communities are sceptical of externally driven innovations, miners thought that technologies had to come from abroad but had to be implemented locally based on the needs of the miners. Government officials stated that some programs had taken place, and officials were sent to advanced countries to learn how mining and land titles had been implemented in other countries. Participants did not take sides as to whether innovation had to be led locally or externally. They stated that it was necessary to implement technologies while bringing all the stakeholders on board, mainly government officials, to "capacitate" them, a word frequently used in the interviews, the miners, and foreign and local consultants. Many also thought that innovation required legal changes, but the participants also believed that digitisation would allow for greater legal compliance.

With lessons from the global terrestrial mining sector, especially in developing countries, ethical and cultural considerations in implementing technology-driven mining solutions highlight the significance of indigenous impetus for innovation. The participants emphasized the need for a local drive towards digitization to address challenges comprehensively, calling for a holistic shift in attitudes and operating systems. This approach extends beyond technological changes, advocating a concomitant shift in the broader cultural context. Lessons from terrestrial mining underscore the ongoing debate about whether innovations should develop indigenously or exogenously. Indigenous innovation (II), often overlooked in discussions of social innovation (SI), has the potential to combine for-profit and non-profit objectives. However, SI practitioners tacitly understand that new social ties initiated by external organizations are essential. The debate reflects concerns about the impact of external interventions on indigenous populations and the potential for unintended consequences. The discussion raises ethical considerations regarding the imposition of external concepts and methods, particularly in participatory development initiatives. The imposition of contemporary democratic systems may undermine communal sovereignty, highlighting the need for a nuanced approach that respects local capacity and existing institutions. The experiences shared caution against disruptive interventions that may overwhelm indigenous populations, emphasizing the importance of community participation in planning and implementation processes. In the mining context, the participants in mining activities in parts of Africa preferred a balanced approach between indigenous and exogenous innovations. The majority supported indigenous innovation, citing concerns that external interventions could complicate existing systems without sufficient support from the mining community. This nuanced perspective aligns with the broader discourse on the necessity of considering local context and engaging stakeholders in implementing technological solutions.

Moreover, the participants emphasized the importance of capacitating all stakeholders, including government officials, miners, and consultants. This collaborative approach seeks to bridge the gap between local implementation and foreign expertise, demonstrating the need for inclusive decision-making processes. The participants acknowledged the role of technology in addressing challenges faced by miners, emphasizing the demand for change within the industry.

13 Overview of Technological, Legal, and Socio-Economic Challenges in Integrating AI and Blockchain Into Space Mining

Developing advanced technologies for prospecting, mining, and processing resources in space poses significant challenges. Autonomous robots, mining equipment, and refining processes must be designed to operate in harsh space conditions. Establishing cost-effective and efficient transportation methods for bringing mined resources back to Earth or other destinations in space is a major logistical challenge.

The economic feasibility of space mining depends on the demand for specific resources on Earth and in space. As technology advances and the need for rare materials grows, the economic case for space mining becomes more compelling. Building the necessary infrastructure in space, such as fuel depots and processing facilities, is crucial for minimizing costs and making space mining economically viable. While space mining is still in its early stages, ongoing research and development efforts are paving the way for a future where the vast resources of our solar system could be harnessed to meet the needs of humanity, both on Earth and in space.

AI (Artificial Intelligence) and blockchain technologies have the potential to significantly enhance the efficiency, transparency, and governance of space mining operations. These technologies can address various challenges associated with space mining, ranging from autonomous decision-making in resource prospecting to ensuring transparent and secure transactions. Here's an overview of their significance:

AI can be employed in autonomous spacecraft and rovers to analyse data from space missions. Machine learning algorithms can identify potential mining sites, assess the composition of celestial bodies, and optimize resource prospecting strategies. This enables more efficient identification of valuable resources in space. Space mining involves complex machinery operating in extreme conditions. AI algorithms can predict maintenance needs by analysing equipment performance data. This proactive approach helps minimize downtime and ensures that mining operations run smoothly.

14 Solutions and Strategic Approaches Derived From Terrestrial Mining Experiences With These Technologies in Extra-Terrestrial Mining

Raising awareness and providing education on blockchain technology within the mining industry can offer numerous benefits, particularly ensuring data immutability and enhancing the efficiency of cadastral systems. Educating miners and state officials on blockchain emphasises the core feature of immutability. This characteristic ensures that once data are recorded on the blockchain, they cannot be altered, providing a secure and tamper-proof record of mining-related information. Highlighting the transparency of blockchain can foster trust among stakeholders. The decentralised nature of blockchain ensures that information is accessible to all authorised parties, thereby reducing the chances of fraud or manipulation. Demonstrating how blockchain technology can streamline processes such as managing mining titles through a digital cadastre can highlight its potential to reduce administrative burdens and enhance overall efficiency. Organising training sessions specifically designed for miners and state officials may be highly effective. These sessions should focus on explaining the fundamentals of blockchain, its application in the mining sector, and its integration with digital cadastre systems. Conducting workshops and seminars that showcase real-world case studies and success stories involving blockchain in mining can help illustrate its practical benefits. Partnering with educational institutions to develop courses or modules on blockchain technology tailored for entry-level miners can be invaluable in raising awareness from an early stage. Utilising various media, such as webinars, online resources, pamphlets, and industry publications, to disseminate information about blockchain and its relevance in mining can reach a broader audience. Training should cover theoretical aspects and provide hands-on experience with digital cadastre systems and blockchain tools relevant to mining title management. Establishing post-training support channels, forums, or helplines can ensure ongoing assistance for individuals who may require further guidance or clarification. The mining industry can gradually enhance its understanding and adoption of blockchain technology by combining awareness campaigns, educational initiatives, and training sessions. This leads to more efficient and transparent mining operations facilitated by secure and immutable data-management systems. Policies should outline clear guidelines for interactions between miners and farmers. This includes delineating boundaries, ensuring compensation for land use, and mitigating potential conflicts over resource utilisation. Introducing blockchain technology to manage mining titles ensures transparency, security, and immutability of records. This instils trust between parties and minimises disputes regarding ownership or rights over mining areas. Using precise geospatial data with a blockchain helps accurately record the coordinates and boundaries. This reduces the possibility of false claims or disputes by providing indisputable evidence for the mining territories. It is important to educate miners, farmers, and relevant stakeholders about the advantages of blockchain to ensure the legitimacy and security of mining titles. This awareness helps build trust and acceptance of the new technology. It is also important to highlight the significance of accurate geospatial data for preventing conflicts and resolving disputes. Communicate how it secures the rights of legitimate owners and prevents encroachment. There is a need to establish standardised protocols for defining specific methodologies and precision levels for recording coordinates. This ensures consistency across all mining operations and minimises the likelihood of conflicting claims due to discrepancies. It is also important to emphasise the importance of precise coordinates in eliminating ambiguities and potential overlaps in mining claims. Uniform standards for accuracy can help to avoid misunderstandings and boundary disputes. It is essential to encourage collaboration among mining authorities, land surveyors, community representatives, and government bodies to ensure inclusive policies and to consider diverse perspectives. It is also essential to monitor adherence to established standards and policies regularly. Implement mechanisms for oversight to promptly address any deviations or discrepancies. Governments and stakeholders can establish a fair and transparent framework by integrating blockchain technology, geospatial information, and standardised pegging practices into miner-farmer relationship policies. This framework minimises conflicts, promotes accountability, and ensures the equitable use of resources in the mining sector while safeguarding the rights of all involved parties.

1. It is crucial to link all the diverse processes of licencing, and title management through blockchain because the benefits of digitisation will not be fully realised without coordination between these different areas.
2. There is a need to visualise mining titles and their coordinates so prospective miners know the areas already being mined and there are no unnecessary conflicts.
3. The precision of the coordinates ensures that the possibility of double allocations is reduced. Blockchain implementation should ensure that the system fails to assign a title to a mine already in possession.
4. Blockchain implementation should also be accompanied by necessary checks that flag any noncompliant behaviour by miners and state officials.
5. It is essential to remap mines using existing maps to introduce discrepancies in digitisation. There is a need to map coordinates with the necessary stakeholders (miners and local communities) again because using physical documents in the digitisation process can result in data manipulation before entering the system.

The implications of the concerns above for space mining are multifaceted, particularly in the context of blockchain technology and its potential integration into the mining industry. Here are several considerations: In space mining, where managing mining titles and coordinates is crucial, blockchain's immutability feature can offer enhanced data security. Once coordinates and mining-related information are recorded on the blockchain, they become secure and tamper-proof, reducing the risk of unauthorized alterations. Transparency is critical in space mining operations, especially when dealing with multiple stakeholders. The decentralized nature of blockchain ensures that information is accessible to all authorized parties, fostering trust. This transparency can be essential when collaborating on space mining projects involving different countries or entities. As the text mentions, streamlining processes through blockchain can significantly enhance efficiency. This is particularly relevant in space mining, where precise coordination and management of mining territories are vital. Implementing blockchain for managing space mining titles can reduce administrative burdens and facilitate smoother operations. Raising awareness and providing education on blockchain technology is crucial for space mining, especially as the industry evolves. Training sessions, workshops, and collaborations with educational institutions can help space miners and officials understand the fundamentals of blockchain and its applications in space resource management. The emphasis on policies outlining clear guidelines for stakeholder interactions directly applies to space mining. With potential collaborations between private space mining companies and governments, clear policies can help manage conflicts and ensure equitable resource use.

In space mining, where precise coordinates are essential, blockchain can help maintain accurate and indisputable records. Highlighting the importance of accurate geospatial data can prevent conflicts and provide clear evidence for mining territories in space. Linking diverse processes through blockchain, visualizing mining titles, and ensuring precision in coordinates are all relevant recommendations for space mining. These measures can help avoid conflicts and streamline licensing and exploration activities. The call for oversight mechanisms to address deviations or discrepancies is crucial in space mining, where compliance with international space laws and agreements is paramount. Implementing blockchain can provide transparency and traceability, aiding oversight and compliance efforts.

15 Analyse Future Technological Trends Impacting Space Mining, Including Quantum Computing and Next-Gen Blockchain Solutions

Quantum computing, as described by Nielsen and Chuang, with its advanced computational power afforded by qubits, poses transformative possibilities for terrestrial mining by addressing complex data challenges and optimising logistics and materials science. Its potential to enhance mineral discovery, streamline planning, and resource allocation cannot be overstated for resource-dependent nations such as Zimbabwe. However, the power of quantum computing also introduces risks, particularly to the cryptographic underpinnings of digital security, in this case, blockchain. Quantum computing is an emerging force that has the potential to revolutionise mineral exploration and data processing but also requires the mining sector to be proactive in safeguarding against disruptive threats through quantum-resistant technologies. The vulnerability highlighted by Aggarwal et al. (2017) suggests that the security of blockchain technologies can be compromised by quantum capabilities, which can rapidly decrypt the current cryptographic safeguards. In response, post-quantum cryptography is burgeoning, aiming to create algorithms that are resistant to the prowess of quantum computing, as detailed by Bernstein and Lange. Incorporating quantum-resistant technologies into blockchain frameworks is essential for Zimbabwe to maintain secure and reliable mining sector applications at the forefront of cyber security.

Integrating quantum computing in Zimbabwe's mining sector holds transformative potential, particularly in addressing complex data challenges and optimizing mineral exploration and resource management. However, it also introduces significant risks concerning blockchain technology, particularly in digital security. Here's an exploration of the implications and considerations: Quantum computing's advanced computational power, harnessed through qubits, offers transformative possibilities for mineral exploration, data processing, and logistical optimization in Zimbabwe's mining sector. The ability to handle complex computations enables more efficient mineral discovery, streamlined planning, and improved resource allocation.

The increased computational power of quantum computing poses a potential threat to the cryptographic foundations of digital security, specifically in the context of blockchain technologies. Traditional cryptographic methods may become vulnerable to quantum capabilities' rapid decryption, as Aggarwal et al. (2017) highlighted. The mining sector must proactively adopt quantum-resistant technologies in response to the identified vulnerabilities. Post-quantum cryptography, as discussed by Bernstein and Lange, involves the development of cryptographic algorithms that can withstand the capabilities of quantum computing. This is crucial for maintaining the security and integrity of blockchain applications in the mining sector.

Incorporating quantum-resistant technologies into blockchain frameworks is essential for ensuring the continued security and reliability of mining sector applications. This proactive approach will help safeguard against potential disruptive threats from quantum computing capabilities, preserving the confidentiality and integrity of mining-related data. Zimbabwe's mining sector, recognizing the transformative potential of quantum computing, should take a leadership role in implementing quantum-resistant technologies. This protects against emerging threats and positions the country at the forefront of cybersecurity within the mining industry.

16 The Role of International Collaboration and Innovation in Addressing the Evolving Challenges of Space Mining

Investment in research and development is crucial for staying ahead of potential cybersecurity challenges posed by quantum computing. Collaborative efforts between the mining sector, research institutions, and cybersecurity experts can contribute to creating robust quantum-resistant algorithms tailored to the unique needs of the mining industry. Given the global nature of quantum computing challenges, international collaboration is vital.

17 Conclusion

Integrating Artificial Intelligence (AI) in space mining operations can revolutionize exploration, extraction, and resource management. AI can analyse vast datasets, predict mineral locations, optimize processes, and enhance decision-making in the extra-terrestrial environment. Blockchain technology provides a secure and transparent framework for space mining operations. It ensures the immutability of data related to mineral exploration, mining rights, and resource allocation. Smart contracts on the blockchain can automate and streamline contractual agreements, enhancing efficiency. Combining AI and blockchain creates a powerful synergy. AI algorithms can process complex data and provide valuable insights, while blockchain ensures the security and transparency of these insights. The decentralized nature of blockchain reduces the risk of fraud and manipulation in space mining operations.

The rise of quantum computing poses security challenges for blockchain. Quantum-resistant technologies are essential to safeguard the cryptographic foundations of digital security in space mining. Addressing these challenges is crucial to maintaining the integrity of extra-terrestrial mining operations. Raising awareness and providing education on AI and blockchain technologies are critical for their successful integration into space mining. Training sessions, workshops, and partnerships with educational institutions can ensure that stakeholders understand and embrace these transformative technologies.

Terrestrial mining operations have accumulated valuable insights, technologies, and best practices. Transferring this knowledge is essential for the successful initiation and sustainable development of space mining endeavours. Lessons learned on Earth can inform and guide space exploration initiatives. Technologies developed for terrestrial mining, such as automation, remote sensing, and AI applications, can be adapted for space mining. Understanding how these technologies have evolved and addressing the unique challenges of space environments can accelerate progress in extra-terrestrial resource extraction.

Insights gained from managing environmental impacts in terrestrial mining can inform responsible practices in space mining. Applying lessons on sustainable resource extraction and minimizing ecological footprints contributes to ethical and environmentally conscious space exploration.

Terrestrial mining has established regulatory frameworks. Extending and adapting these frameworks to space mining ensures that legal and ethical considerations are addressed. Collaboration between space agencies, governments, and international organizations is crucial for effective regulation. Collaboration between terrestrial mining industries, space agencies, research institutions, and private enterprises can facilitate knowledge exchange and joint initiatives. Sharing expertise, technologies, and resources accelerates progress in terrestrial and extra-terrestrial mining endeavours.

In conclusion, integrating AI and blockchain holds immense promise for space mining operations, offering efficiency, security, and transparency. Leveraging insights from terrestrial mining is pivotal for a responsible and sustainable approach to space exploration, ensuring extra-terrestrial mining activities' ethical and successful development. Collaborative efforts and proactive measures are vital for unlocking the full potential of these advancements in the emerging frontier of space mining.

References

Aarvik, P. (2020). Blockchain as an Anti-Corruption Tool.

Abbud-Madrid, A. (2021, June 28). Space Resource Utilization. Oxford Research Encyclopedia of Planetary Science. Retrieved March 15, 2024, from https://oxfordre.com/planetaryscience/view/10.1093/acrefore/9780190647926.001.0001/acrefore-9780190647926-e-13

Abbud-Madrid, A., Beaty, D. W., Boucher, D., Bussey, B., Davis, R., Gertsch, L., Hays, L. E., Kleinhenz, J., Meyer, M. A., Moats, M., Mueller, R. P., Paz, A., Suzuki, N., van Susante, P., Whetsel, C., & Zbinden, E. A. (2016, April). Report of the Mars Water In-Situ Resource Utilization (ISRU) Planning (M-WIP) Study (p. 90).

Abramov, V., Sikarev, I., Chusov, A., Tatarnikova, T., & Shilin, M. (2021). Digital Transformation of Expert Systems Design Within Geoinformation Management in Arctic. *Journal of Physics: Conference Series, 2001*, 012014. https://doi.org/10.1088/1742-6596/2001/1/012014

Aditya, T., Santosa, P. B., Yulaikhah, Y., Widjajanti, N., Atunggal, D., & Sulistyawati, M. (2021). Title Validation and Collaborative Mapping to Accelerate Quality Assurance of Land Registration. *Land Use Policy, 109*,. https://doi.org/10.1016/j.landusepol.2021.105689

Aggarwal, D., Brennen, G. K., Lee, T., Santha, M., & Tomamichel, M. (2017). Quantum Attacks on Bitcoin, and How to Protect Against Them. arXiv:1710.10377.

Alam, K. M., Ashfiqur Rahman, J. M., Tasnim, A., & Akther, A. (2020). A Blockchain-based Land Title Management System for Bangladesh. *Journal of King Saud University, Computer and Information Sciences.*

Asiama, S. O. (2003). Comparative Study of Land Administration Systems: Case Study of Ghana (Peer Review of World Bank Policy Research Report). https://www.google.com.gh/search?q=Comparative+study+of+land+administration+systems:+Case+study+of+Ghana+pdf&ie=utf-8&oe=utf-8&gws_rd=cr&ei=9MYOV6fsNMmKsgGfwrnYCgF

Asp, P. J. (2000). Mining and Indigenous Peoples (special issue). UNEP's Industry and Environment 63.

Badescu, V., & Zacny, K. (2018). *Outer Solar System: Prospective Energy and Material Resources.* Springer.

Benjamin, S., & Raman, B. (2011). Illegible Claims, Legal Titles, and the Worlding of Bangalore. *Dans Revue Tiers Monde, 2*(206), 37–54.

Benjamin, S., Bhuvaneswari, R., & Rajan, P. (2007). Bhoomi: 'E–Governance', or, an Anti-Politics Machine Necessary to Globalize Bangalore? (CASUM–m Working Paper).

Bennett, R., Miller, T., Pickering, M., & Kara, A.-K. (2021). Hybrid Approaches for Smart Contracts in Land Administration: Lessons from Three Blockchain Proofs-of-Concept. *Land, 10*, 220. https://doi.org/10.3390/land10020220

Bonnet, D., & Nandan, P. (2011). Transform to the Power of Digital—Digital Transformation as a Driver of Corporate Performance. Available via Capgemini Consulting. https://www.capgemini.com/resources/transform-to-the--power-of-digital. Accessed 4 April 2016.

Castro, A., & Nielsen, E. (2001). Indigenous People and Co-management: Implications for Conflict Management. *Environmental Science & Policy, 4*, 229–239. https://doi.org/10.1016/S1462-9011(01)00022-3

Christensen, I., Lange, I., Sowers, G., Abbud-Madrid, A., & Bazilian, M. D. (2019). New Policies Needed to Advance Space Mining. *Issues in Science and Technology*, *35*(2), 26–30. https://www.jstor.org/stable/26948985

Dale, P., & McLaughlin, J. (2011). Land Administration. https://doi.org/10.1007/978-94-007-1667-4_15.

Deshpande, A. (2019). An Inquiry into the Impact of Digitization and Customized ERP Applications on Twin Engineers Overall Efficiency—An Empirical Study Approach.

Desmet, D., Duncan, E., Scanlan, J., & Singer, M. (2015, September). Six Building Blocks for Creating a High-Performance Digital Enterprise. McKinsey Digital. McKinsey & Company. http://www.mckinsey.com/business-functions/organization/our-insights/six-building-blocks for-creating-a-high-performing-digital-enterprise. Accessed 22 January 2016.

Dobrolyubova, E., Klochkova, E., & Alexandrov, O. (2019). Digitalization and Effective Government: What is the Cause and What is the Effect? *International Conference on Digital Transformation and Global Society* (pp. 55–67). Springer.

Dorian, J. P., & Humphreys, H. B. (1994). Economic Impacts of Mining. *Natural Resources Forum, 18*, 17–29. https://doi.org/10.1111/j.1477-8947.1994.tb00869.x

Dorner, K., & Meffert, J. (2015, October). Nine Questions to Help You Get Your Digital Transformation Right. Mckinsey Digital. McKinsey & Company. http://www.mckinsey.com/businessfunctions/organization/our-insights/nine-questions-to-help-you-get-your-digital-transformation-right. Accessed 22 January 2016.

Durcikova, A., Fadel, K. J., Butler, B. S., & Galletta, D. F. (2011). Knowledge Exploration and Exploitation: The Impacts of Psychological Climate and Knowledge Management System Access. *Information Systems Research, 22*(4), 855–866.

Evans, O. (2018). Digital government: ICT and Public Sector Management in Africa (MPRA Paper, No. 91628) (pp. 269–286).

Ferdous, M. S., Chowdhury, F., & Alassafi, M. O. (2019). In Search of Self-Sovereign Identity Leveraging Blockchain Technology. *IEEE Access, 7*, 103059–103079. https://doi.org/10.1109/access.2019.2931173

Gimpel, H., & Roglinger, M. (2015, October). Digital Transformation: Changes and Chances—Insights Based on an Empirical Study. Project Group Business and Information Systems Engineering (BISE) of the Fraunhofer Institute for Applied Information Technology FIT.

Glaser, F. (2017). *Pervasive Decentralisation of Digital Infrastructures: A Framework for Blockchain Enabled System and Use Case Analysis*. HICSS.

Goswami, B., Jakhu, R., & Chen, D. (2020). Threats to Peaceful Purposes of Outer Space: Politics and Law.

Gupta, A. K. (2013). Tapping the Entrepreneurial Potential of Grassroots Innovation. *Stanford Social Innovation Review*, 18–20

Hilson, G. (2002). The Environmental Impact of Small-Scale Gold Mining in Ghana: Identifying Problems and Possible Solutions. *The Geographical Journal, 168*(1), 57–72.

Hornsey, M. J., Fielding, K. S., Harris, E. A., Bain, P. G., Grice, T., & Chapman, C. M. (2022). Protecting the Planet or Destroying the Universe? Understanding Reactions to Space Mining. *Sustainability, 14*(7), 4119. https://doi.org/10.3390/su14074119

Huber, G. P. (2001). Transfer of Knowledge in Knowledge Management Systems: Unexplored Issues and Suggested Studies. *European Journal of Information Systems, 10*(2), 72–79.

Hull, M. S. (2012). *Government of Paper: The Materiality of Bureaucracy in Urban Pakistan* (320 pages). University of California Press.

Iansiti, M., & Lakhani, K. (2017). The Truth about Blockchain. *Harvard Business Review*, 1–11. https://enterprisersproject.com/sites/default/files/the_truth_about_blockchain.pdf

Joshi, K. D., Lei, C., Datta, A., & Shu, H. (2010). Changing the Competitive Landscape: Continuous Innovation Through IT-Enabled Knowledge Capabilities. *Information Systems Research, 21*(3), 472–495.

Kent, C. A. (2019). Surface and Subsurface Property Rights in Minerals: The Continuing Conflict. *The Journal of Private Enterprise, 34*(2), 43–64.

Knirsch, F., Unterweger, A., & Engel, D. (2019). Implementing a Blockchain From Scratch: Why, How, and What We Learned. *EURASIP Journal on Information Security, 2019*,. https://doi.org/10.1186/s13635-019-0085-3

Konashevych, O., & Cross-Blockchain Protocol for Public Registries. (2020, April 24). The Paper is Accepted for Publication in the International Journal of Web Information Systems, Emerald Publishing. This is a Pre-print Version under Creative Commons Attribution Non-commercial International Licence 4.0 (CC BY-NC 4.0). https://doi.org/10.1108/IJWIS-07-2020-0045, https://ssrn.com/abstract=3537258 or https://doi.org/10.2139/ssrn.3537258

Lange, S. (2008). Land Tenure and Mining in Tanzania.

Lemieux, A. (1997). Canada's Global Mining Presence. In *Canadian Minerals Yearbook*. Natural Resources Canada.

Lührmann, A., Marquardt, K. L., & Mechkova, V. (2020). Constraining Governments: New Indices of Vertical, Horizontal, and Diagonal Accountability. *American Political Science Review, 114*(3), 811–820.
Marwick, A. E. (2013). *Status Update: Celebrity, Publicity, and Branding in the Social Media Age* (pp. 1–36).
Mbendi. (1999). Information for African Mining Statistics. Africa's Leading Business Website. http://www.mbendi.co.za/indy/ming/mingaf.htm#Overview.
McLeod, H. (2000). Compensation for Landowners Affected by Mineral Development: The Fijian Experience. *Resources Policy, 26*, 115–125.
Meijer, A. (2006). ICTs and Political Accountability: An Assessment of the Impact of Digitization in Government on Political Accountability in Connecticut, Massachusetts and New York State (NCDG Working Paper, 17).
MiningWatch. (2000). On the Ground Research: A Workshop to Identify the Research Needs of Communities Affected by LargeScale Mining. MiningWatch Canada and the Canadian Consortium for International Social Development (CCISD).
Moreri, K. K., Fairbairn, D., & James, P. (2018). Volunteered Geographic Information Quality Assessment Using Trust and Reputation Modelling in Land Administration Systems in Developing Countries. *International Journal of Geographical Information Science, 32*(5), 931–959.
Mougayar, W. (2016). *The Business Blockchain: Promise, Practice, and Application of the Next Internet Technology*. Wiley.
Nakamoto, S. (2008). Bitcoin: A Peer-to-Peer Electronic Cash System. https://bitcoin.org/bitcoin.pdf
Odhiambo, M. O. (2000). *Oxfam Karamoja ConflictStudy: A Report*. Oxfam.
Olatokun, W. M. (2008). Gender and National ICT Policy in Africa: Issues, Strategies, and Policy Options. *Information Development, 24*(1), 53–65.
Ølnes, S., Ubacht, J., & Janssen, M. (2017). Blockchain in Government: Benefits and Implications of Distributed Ledger Technology for Information Sharing. *Government Information Quarterly, 34*, 3.
Paek, S. W., de Weck, O., Hoffman, J., Binzel, R., & Miller, D. (2019). Optimization and Decision-making Framework for Multi-staged Asteroid Deflection Campaigns under Epistemic Uncertainties. *Acta Astronautica, 167*,. https://doi.org/10.1016/j.actaastro.2019.10.042
Resiere, D., Resiere, D., & Kallel, H. (2020). Implementation of Medical and Scientific Cooperation in the Caribbean using Blockchain Technology in Coronavirus (Covid-19) Pandemics. *Journal of Medical Systems, 44*, 123. https://doi.org/10.1007/s10916-020-01589-4
Ryder, S. S., & Hall, P. M. (2017). This Land is Your Land, Maybe: A Historical Institutionalist Analysis for Contextualizing Split Estate Conflicts in U.S. Unconventional Oil and Gas Development. *Land Use Policy, 63*, 149–159.
Sedlmeir, J., Smethurst, R., Rieger, A., et al. (2021). Digital Identities and Verifiable Credentials. *Business & Information Systems Engineering, 63*, 603–613. https://doi.org/10.1007/s12599-021-00722-y
Sercel, J., Peterson, C., Britt, D., Dreyer, C., Jedicke, R., Love, S., & Walton, O. (2018). Practical Applications of Asteroidal ISRU in Support of Human Exploration. https://doi.org/10.1016/B978-0-12-813325-5.00009-4
Siddiquee, N. A. (2005). Public Accountability in Malaysia: Challenges and Critical Concerns. *International Journal of Public Administration, 28*(1/2), 107–129.
Sonter, M. J. (1998, September 28–October 2). Presented at 49th International Astronautical Federation Congress.
Tchamyou, V. S., Erreygers, G., & Cassimon, D. (2019). Inequality, ICT and Financial Access in Africa. *Technological Forecasting and Social Change, 139*, 169–184.
Thontteh, E. O., & Omirin, M. M. (2015). Land Registration within the Framework of Land Administration Reform in Lagos State. *Pacific Rim Property Research Journal, 21*(2), 161–177. https://doi.org/10.1080/14445921.2015.1058033

Wade, M. (2015). Digital Business Transformation: A Conceptual Framework. Global Center for Digital Business Transformation—An IMD and Cisco Initiative. http://www.imd.org/uupload/IMD.WebSite/DBT/Digital%20Business%20Transformation%20Framework.pdf. Accessed 22 January 2016.
Xie, R., Saydam, S., & Dempster, A. (2022). A Framework for Campaign Level Asteroid Mining Pre-Feasibility Study. https://doi.org/10.2514/6.2022-2583.

An Investigation into Counter Drones Digital Forensics Aided by Remote Identification

Lacramioara Nicoleta Ivascu and Hamid Jahankhani

Abstract While drones are presenting numerous benefits, they also could be misused by nefarious actors involved in incidents caused by airspace intrusion in restricted areas, unauthorised surveillance, smuggling contraband, espionage, cyber-attacks (GPS spoofing, malware injection, denial of service DoS, video feed manipulation, wardriving, physical tampering, naval piracy etc.), weaponization etc. Counter drones are crucial, but they don't solve all the airspace traffic problems. To address holistically the technical, legal, and political challenges posed by drones and enhance the overall airspace security, aviation authorities around the globe have taken various actions for introducing regulatory measures, some of them being based on introducing the remote identification (RID) for drones. The UK is taking steps towards assessing, and if agreed, implementing an RID system. RID enables the in-flight real-time identification, showing information referring to the drone's serial number, location, the pilot's identity etc. This chapter aims to offer an insight of the role of RID in the drone digital forensics and countering rogue drones, by combining knowledge of drones, counter-drones, digital forensics, and remote identification and showing how regulatory measures and technology can work together to deliver responsible and secure use of drones in our continuously developing world.

Keywords Counter drones · Digital forensics · Remote identification · RID · Cyber attack · Intrusion detection · GPS spoofing

1 Introduction

The humans have always looked with curiosity to the sky, trying to fly and explore new horizons. From the very first crafted wings, imitating birds and insects to the most powerful 6th generation hypersonic nuclear-powered jets, sophisticated stealth unmanned aerial systems (UAS) and flying cars of today the journey in time has rapidly accelerated in the last years, being facilitated by the rapid advancements

L. N. Ivascu · H. Jahankhani (✉)
Northumbria University, London, UK
e-mail: Hamid.jahankhani@northumbria.ac.uk

H. Jahankhani et al. (eds.), *Space Governance*, Space Law and Policy,
https://doi.org/10.1007/978-3-031-62228-1_7

in technology, adaptability and affordability but also being motivated by the latest geopolitical developments in world. Unmanned aerial systems are representing one of the newest fields in science and it seems that they are revolutionising the way our daily lives are going to look in the near future.

The proliferation of unmanned aerial vehicles across various industries, either civilian or military, combined with the rise in security threats, are expanding at a faster pace than the regulatory landscape can keep up with. Digital forensics need to be modernised and integrated with counter drones technologies, in a collaborative international effort, given the actual geo-political state of the world. Integrating the Remote Identification with the drones ecosystem has the potential to enhance the counter drones ability to identify unmanned aerial vehicles, and if found as being malicious, the next step would be to involve the digital forensics technologies into the chain of custody.

The chapter aims to investigate how implementing the remote identification for drones would aid the counter-drones' abilities to detect, identify, monitor, isolate, annihilate nefarious UAVs, as part of the UK's airspace modernisation strategy and how the digital forensics for drones and counter-drones will respond to the implementation of the remote ID.

2 Literature Review

2.1 *Drones*

According to the Cambridge Dictionary definition, a drone is 'an aircraft that does not have a pilot but is controlled by someone on the ground, used especially for dropping bombs or for surveillance (=careful watching of a place)'. This definition seems somehow overtaken by the many uses drones could have, 237 of them being described by Alvarado, 2021.

PWC has predicted for 2030 the UK's total population of drones in use by commercial organisations and government could be more than 920,000 and £45 billion by 2030, of which more than a fifth could be used by the public sector and the potential annual cost savings from drone adoption in 2030 will increase by £22.4billion.

Drones can be classified by their build as fixed-wing, rotary-blade, a hybrid of both; by their environment as active in the air, underwater or amphibious; by their way of communication as controlled remotely or use satellite navigation systems or other methods to fly automated, semi-autonomously or autonomously.

The main components of a drone are:

- Hardware

The physical components are mainly: drone body, flight controller, motors, rotors/propellers/wings, electronic speed controllers (ESC), protective casing, position sensors (GPS Receiver, Magnetometer or Compass, Altimeter), Inertial measurement Unit (IMU) and related sensors, Radio Receiver (RX), Transmitter (TX), LED Lights, fixed wings specific components (launching and landing mechanisms etc.). Other components are: Remote controller, Mobile/tablet device, First person View (FPV) Goggles, Memory cards, batteries, Cloud storage etc.

- Software

Their software is formed out of the Flight Management Software and Ground Control Software. They control the Ground Control Station (GCS) and the Command and Control Link (C2 Link, or the 'radio line of sight').

- Data

The Drone Data could be classified as: Audio Visual Content, Flight Schedules, Other Payload Created Content, Automated Usage Logs etc. Drone data can be stored on board of the drone, on removable Storage devices, on mobile devices and applications, on remote controllers, on ground stations, on cloud-based data platforms but also as network packet data.

- Payload

The drone payloads could be referring to camera and video equipment, delivery payload, weapons, fuel tanks, communications devices, humans, other drones etc.

- Communication

A drone receives from the Ground Control Station (GCS) movement commands from control sticks, other controls commands such as Return to Home (RTH), camera controls and it transmits back to the GCS motion telemetry (position, altitude, orientation, velocity etc.), health telemetry (battery life remaining) and camera imagery.

The flight controller is the brain of the drone, driving signals to the motors to implement the pilot's movement requests, while it's compensating for environmental factors (wind etc.). It also has other functions as camera control, special flight modes, obstacle avoidance, battery monitoring. The flight controller needs to understand the current position (altitude, latitude, longitude), orientation, current velocity (direction and speed).

The Global Navigation Satellite System (GNSS) is a generic term for various satellite navigation systems, the main three ones being GPS (USA, launched in 1978), GLONASS (Russia, launched in 1982 and again in 2010), Galileo (EU, launched in 2016) operating globally, BeiDou (China), NavIC (India) and QZSS (Japan), operating locally. If two or more of these technologies are combined, the result is a faster and more accurate location as there is a higher range of satellites available and they are specialised for various environments (urban-Galileo, mountains-GLONASS, rural-GPS).

- Power

Drones could be powered by conventional Lithium batteries, solar paint/batteries, hydrogen, and even nuclear fusion.

2.2 Drones in the Sky

Drones could be extremely useful either in the civilian or the military fields. Based on empiric observation of the drones evolution over the last few years, it is likely that the humans curiosity and ingenuity will find more uses for the drones. Here are some examples of drones uses known so far:

- Police and border law enforcement, security and surveillance of infrastructure objectives and prisons.
- Medical, used for delivering medicines and instruments (example: defibrillators), collecting tests samples, transporting emergency transplant organs, transporting patients and also injured soldiers in the war zones. In Europe the longest route for drone delivery for medical purposes is implemented by MedLife in Romania, across 120 km long route.
- Environment conservation, combating animal poaching, wildlife conservation, forestry, livestock monitoring, marine life management; Meteorology and natural phenomena; Agriculture, aquaculture, mariculture—inspection, irrigations, animal fare, pollination; Detect cannabis farms.
- Constructions and Infrastructure—inspection, repairs, defrost, painting, cleaning etc. of buildings, railways, roads, wind turbines, petrol rigs, electricity posts, pipelines, power lines, refineries, bridges, tunnels etc.; Solar farms inspection and maintenance, repairing fissured or broken solar panels; Aircraft inspection and maintenance.
- Aerial 2D and 3D mapping, in the air, underground or underwater.; Archaeology and forensics, used to unravel sites of interest underground and underwater; Real estate photography and planning inspections; Smart traffic and parking monitoring.
- Disaster response, firefighting, water or fire rescue, emergency lighting.
- Transportation and logistics—taxi, deliveries of goods; Warehouse logistics and stock management.
- Relay communication; Internet access in areas with week or no signaL; Follow me used for guiding people and other devices.
- Re-fuelling fight jets (stealthy Boeing MQ25 Stingray).
- Drone racing; Entertainment, photography, videography, concert shows, New Year drones displays, Formula1 race track signage; toys for all ages.

The engineers and hobbyists' creativity is also reflected in the variety of the materials used, from the most rudimentary ones to the most advanced stealth materials. Some examples are cardboard assembly packs sent from Australia to Ukraine during the

Russia-Ukraine war., 3D printed components mainly for racing drones, rice cakes/ vegetables and gelatine for emergency nutrients for stranded people, recycled sky jets components for underwater kamikaze drones. etc.

Other impressive UASs are:

- Lockheed Martin RQ-170 Sentinel (US, CIA)
- Sukhoi S-70 Okhotnik-B (RU)
- Northrop Grumman X-47B (US)
- Boeing Phantom Ray (US)
- Dassault nEUROn (FR)
- BAE Systems Taranis (UK)
- Chengdu J-20 (CN)
- Shenyang J-31 (CN)
- General Atomics Avenger (US)
- General Atomics MQ-9 Reaper (US)
- DRDO Ghatak (IN)
- Kratos XQ-58 Valkyrie (US)
- Hongdu GJ-11 sharp sword (CN)

2.3 Drones Underwater

Underwater Unmanned Vechicles (UUVs) could be Remotely Operated Underwater Vehicles (ROUVs) or Autonomous Under Water Vehicles (AUVs) and Autonomous and Remotely Operated vehicles (ARV).

The ROUVs are manually controlled by an operator and they are very useful in replacing humas for difficult tasks underwater. One of the largest ROUVs can rescue up to 16 people in one go inside a Submarine Rescue Diving Recompression System (SRDRS).

The AUVs range from just a few kilograms to screw driven AUVs, Bionic AUVs and underwater gliders. Deepglider, one the most recent AUVs, weights just over 60 kg, can travel 8500 km and dives as deep as 6000 m.

The designers of UUVs are trying to improve the quality of the movement underwater (glider design), the stealthiness (manta ray shape) and the ability to be self-sufficient regarding the power sources, so it can stay underwater for longer periods of time (oxygen/hydrogen air -independent propulsion and lithium and water power source).

UUVs are used for disabling underwater mines or for intelligence, reconnaissance and submarine warfare, as shown in the survey conducted by RAND Corporation. Some underwater drones are capable of air travel too, being very efficient in performing military tasks.

Other uses are related to cinematography, deep-sea exploration and research such as measuring micro-plastic contents of the water, temperature, currents, 3D sea bed mapping, deep-sea animal and plants observation etc.

The challenges faced by the UVVs are related to their communication with the pilot, as they use acoustic waves rather than electromagnetic waves and those are more vulnerable due to the absorption of the signal, reflection and refraction.

UUVs are difficult to counter as they can easily avoid detection underwater. Bai et al. (2020) have published a survey about UUVS, listing 'the sensing and recognition systems' and classifying them in acoustic-sensors (sonar) and sound waves (hydrophones); optical sensors (using camera and light detection), light waves (LiDAR), chemical sensors, physical sensors (Bai et al., 2020).

Remote identification is not yet made available for UUVs but, as the underwater activity is increasing, RID could be implemented underwater in symbiosis with other systems, not far in the future.

2.4 Drones Underground

Drones are not only active in the air or underwater, but also in the underground. Initially developed by Exyn EI Technologies for high risk environments, such as mines explorations and 3D mapping, the ExynAero drones operate autonomously (Level 4, no human presence required) beyond line of sight (BLOS), without GPS, Wi-Fi, landmarks, pre-loaded data or pilot, by exploring sensing modalities as LiDAR, inertial sensors and own SLAM pipeline, avoiding collision and adapting to dust and degraded lighting. Anecdotally, during a product demo in 2020 the Start button was pressed by a dog and the drone completed the task successfully without any human intervention.

Nowadays they are used in the war, in the tunnels of Gaza.

Another AI driven small drone is Nova 2, made by Shield AI, used to map a multi-story and underground building before the soldiers access it. Nova 2 not only that it creates sophisticated 3D maps of the environment, but it also maps the humans present, so the soldiers are always aware of who is in the building or tunnel.

- **Swarms**

What is better than just one autonomous AI driven drone? Two or more, in an 'adaptive, complex swarm, with collective behaviours for intelligent movement, decisions, and interactions with the environment' fully autonomous drones working together, in a platform with an open architecture, so any number or other drones and robots could be added. Elbit Systems developed this platform and it incorporates its Legion-X system, but also Lanius and Probot, for combined search and attack capabilities.

In the beginning 'the bigger the better' was true and the hyper drones were impressive, showing off signs of power. Nowadays they are still popular but they are getting outshined by the medium and smaller size drones, the main reason being financial. The smaller drones are easier, cheaper, faster to produce, to deploy, to pilot, and to replace in the war situations. There are many abandoned projects for building large drones either because they are too costly or because their technology is already obsolete, being eclipsed by smaller, more efficient drones.

In his book 'Swarm Troopers, How small drones will conquer the world', David Hambling highlights the advantages of small drones swarms given by their capacity to extract their 'energy from the environment with solar cells, dynamic soaring and power scavenging', by imitating the nature, flying like a swarm of bees, while being able to 'slice through steel and disintegrate concrete'. Small drones are just smart phones with wings, and the wings are the cheap components (Hambling, 2015).

There are various types of swarms, from the 'classic' ones where they could be made of a collection of similar drones following a lead drone, that is controlled by a pilot, to swarms formed out of drones, robots and other IoT devices, driven by AI, with various degrees of autonomy.

Swarm challenges are still found in communication protocols, battery capabilities and network layers. The drone entertainment show we watch at various occasions are not real swarm technology, but they are centrally choreographed.

If securing single drones is proving being a complex process, securing drones swarms becomes even more complicated. In the paper 'A Distributed Agent-Based Framework for a Constellation of Drones in a Military Operation' the authors propose a successfully simulated framework for secure communications between drones in a constellation, based on peer-to-peer (P2P) and client–server (C/S) connectivity by using identity authentication (Fitwi et al., 2019).

In their paper, Omar Shrit et al. present their tested solution to synchronise and orchestrate a swarm of drones by using autonomous drones, connected via Wi-Fi and following a leader drone that is controlled by a human pilot (Shrit et al., 2017).

Swarms could be powerful and combined with AI they have an unknown potential, with disastrous consequences. Hence the US military plans to create huge autonomous drone swarms spark concerns. One of these projects, the AMASS would involve thousands of drones, on the ground, in the air and in the water, working together in a "swarm of swarms" to overwhelm enemy defences, as David Hambling presents in his article. Another project is 'The Replicator' that is focused on the underwater swarm, in order to create the 'silent swarm' as response to the China's plans to develop AI-driven swarms formed out of tens of thousands of drones.

In 'A Review of Security Incidents and Defence Techniques Relating to the Malicious Use of Small Unmanned Aerial Systems' the authors present their review of malicious drones incidents around significant objectives as nuclear facilities and airports. The authors argue that although there are successful counter measure available on the market, the policy and legal challenges are a step behind (Swinney & Woods, 2022).

Another analysis of concerns related to UAS security and privacy with their recommendations and countermeasures, but also to the regulations at the moment of writing is presented in 'Analysis on security-related concerns of unmanned aerial vehicle: attacks, limitations, and recommendations' (Siddiqi et al., 2022).

- **Drone attacks**

As useful as they could be, drones are a double-edged sword as they could also be used for malicious purposes, such as surveillance, invasion of privacy, sabotage, damage to industrial sites, smuggling, contraband, espionage, and intelligence gathering, air traffic disruptions, cyberattacks, harassment and threats, collisions, weaponization etc. some of these having disastrous consequences.

In the paper 'A comprehensive Survey: Cybersecurity Challenges and Futures of Autonomous Drones' the researchers show that drones are vulnerable to cyber attacks but they also could pose serious cyber threats. Drones are not sufficiently protected against threats regarding authentication, signal verification, encryption, access control, threat reporting. Mobility and stealth are qualities that drones need in both offence and defence. Therefore, detection is important because if the drones are detected, they could then be identified, captured, diverted, annihilated. Many countries and international law enforcing agencies have produced laws designed to reduce the activities of nefarious actors or to ban them completely (Durfey & Sajal, 2022).

Drones could be attacked and hacked in the same ways as any other IoT devices, by using Wi-Fi attacks, firmware reverse engineering, GPS spoofing/jamming etc. and they are used to attack other devices too. Some of the notorious examples of drones incidents are:

- 2011—The CIA drone RQ-170 Sentinel was hacked and taken down while flying over Iran, its remains being captured by the Iranian forces.
- 2013—A drone of US Department of Homeland security was hacked by a security researcher.
- 2015—'SkyJack' created a program that could turn drones into zombies for malicious purposes.
- 2105—In Japan an environmental activist, Yasuo Yamamoto, strapped a container with radiations contaminated sand from Fukoshima on a small drone and landed it on the roof of the Prime minister house to radiate for two weeks, until it got found by accident ('Drone "containing radiation" lands on roof of Japanese PM's office').
- 2015—'A Drone, Too Small for Radar to Detect, Rattles the White House', crash-landing on its lawn.
- 2016—An ethical hacker could hack a DJI drone via a firmware vulnerability.
- 2017—Apple declares its campus a 'No drone zone' and Tesla and Meta had incidents of attempts to drone espionage too.
- 2018—A hacker got into the software of a Parrot (Bouafif et al., 2020) by using a $40 tool. In the same year the Heliguy manufacturer issued a warning related to the security of drones, after high-profile drones caused disruption to Gatwick airport.
- 2018—in Venezuela there was an assassination attempt performed by two drones that detonated above the president's head.
- 2019—Researchers from the University of Birmingham were able to exploit the GPS vulnerabilities of a commercial drone, causing it to fly erratically (Parkinson et al., 2017).

- 2022—hackers used Wi-Fi drones to penetrate a financial firm's network remotely by dropping two modified drones, equipped with a pen kit, on the roof of the building.

Some of these attacks have been performed by using commercial, inexpensive drones. In his report, Short (2019) shows how the law enforcement agencies and military forces are paying increased attention to the misuse of the Commercial-of-the-Shelf (COTS) drones in the vicinity of strategic buildings, military objectives, airports, prisons, public events etc. (Short, 2019).

Niyonsaba and his colleagues present in their paper a taxonomy of UAS drones attacks and a survey of defence technics, discussing Machine Learning and other future research suggestions (Niyonsaba et al., 2023).

The paper 'A Comprehensive Study of Various Techniques for Hostile Drone Detection and Their Classification' presents a classification of drones based on size, features and flight height. Unique Identification Number gets discussed in this paper (Dani et al., 2021).

In the paper 'A Security Review in the UAVNet Era: Threats, Countermeasures, and Gap Analysis' the authors present a review of drones security threats and their traditional or modern countermeasures and perform a gap analysis of the emerging ad-hoc and also cloud-based infrastructures (Rugo et al., 2023).

In A Survey on Detection, Tracking, and Classification of Aerial Threats using Radars and Communications Systems' the authors present an overview of various radar systems, one of them being the Joint Communication-radar (JCR) system, comparing them with other systems that do not involve radar for in their missions to detect, track and classify drones threats (Khawaja et al., 2022).

A few decades ago, the knowledge for technological developments in drones was reserved to the military and academia. With the proliferation of the Internet it became more affordable and accessible to the civilians. Nowadays the military is taking an increasing interest in the warfare. The face of the war is changing rapidly and, sadly, it is a growing business. Two of the growing branches of the modern war are the private security (mercenaries) groups and the cyber hacker organisations. Many well-known international mercenary groups are becoming better trained and equipped, hiding under the private security businesses names like Blackwater, Academi, Triple Canopy, Constellis Group, Apollo Global Management etc., aiming for high-risk high reward activities. In the US there are 1.8 milion people active in this sector. In Russia, the Wagner group is well-known for its deployment in Syria and Africa. The other group are the hackers who, just like the mercenaries, are employed by well organised businesses. If 20–30 years ago hackers were mostly lone actors and they were part of the punk culture, nowadays they are on a payroll and get fired if they are late on work, hacking being a serious business.

Both these groups are becoming better trained and equipped, internationally active, but also disposable. While their casualties don't return to their countries wrapped in the national flag, these groups could cause a lot of damage if they get access to advanced war technologies, including drones and counter-drones. Sky is not the limit anymore, but how many drones they could buy, and drones are becoming cheaper to make. The Turkish, Iranian drones and missiles are more affordable. Why spend tens of millions on one Reaper when one could buy so many Baykatars for $20k each? The DJI drones are very popular in Ukraine as they cost under $500 and could destroy an expensive tank. The Ukrainian army engineers and volunteers are modifying these drones to be able to avoid the Russian counter drones and carry ammunition payloads.

In the same time Russian army improvise and experiment with modified drones too. At the time of writing this paper Russia has sent 75 of what it seemed to be Shahed136-type to attack Ukraine. They were repainted with a stealth material and some of these drones had a 4G Cellular Modem with SIM card from an Ukrainian telecom provider, with antennas, mounted on tape. Although the electronic signature increases by adding cellular communication to a drone, this way the Russians could get real-time intelligence whether the drones were successful or not in their attack, and maybe even where the counter drones systems were.

Due to the restless geopolitical situation evolving in various areas of the globe, the development of expensive, advanced, sophisticated drones has been on the agenda of many countries. At the moment more than 100 countries own military drones and their number is increasing. There are many intricate political affairs regarding procurement of drones and drone parts, many countries trying to restrict or block their opponents' supplies channels. Other countries are importing drone parts via their neighbouring countries.

In the beginning of the Ukraine-Russia war, while the Ukrainians were using lots of DJI drones for their attacks, they noticed that often if they would launch a drone, shortly after that the take off location would be hit back by the Russians. That was happening because the Russians used to purchase via Syria the Aeroscope device that would help decode the DJI DroneID data (Fig. 1).

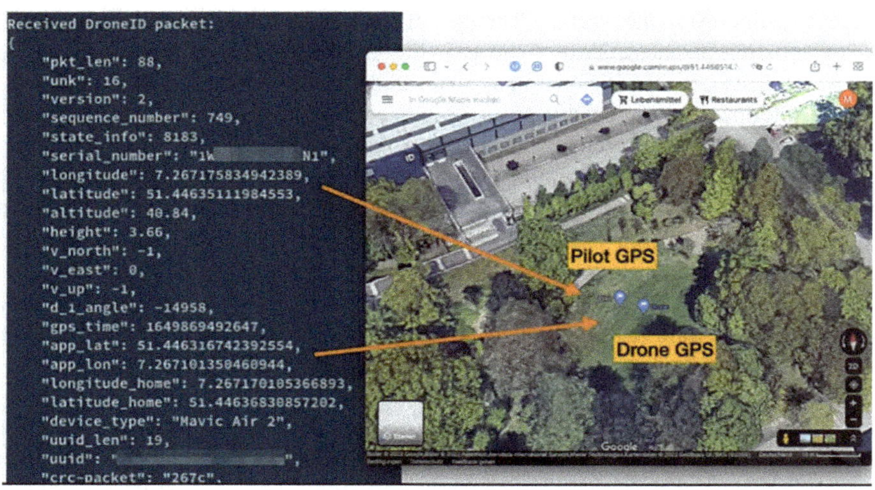

Fig. 1 How DJI drones show the location of the drone and of the pilot

DJI, the same company that admitted lying in 2022 about the DroneID data being encrypted, stopped selling to both countries. That led to a shortage of small drones on the frontline. The News about contraband with drones and drone parts are increasing, for instance how Finland captured a group that was sending illegally tens of thousands of drones to Russia.

In the meantime, more improved and affordable types of drones get developed. On 19th of November the newer Shahed 238 was displayed, this version being jet-powered and 'with radar and electro-optical/infrared guidance', so more devastating for Ukraine if they get purchased and used by Russia.

- **Counter Drones**

Undoubtedly the need for drones, either used for civilian or military purposes, results in a natural and at least equally significant need for counter-drones. Because drones are easy to acquire, especially the commercial ones, and the malicious actors are becoming more creative, the counter-drone industry needs to be one step ahead of the game to be able to perform detection, identification, and mitigation, in order to reduce the attack risk and improve the safety and security of the population. There is not a simple solution to cover all drone issues, but effective counter drone technologies combined with the right regulation are the way forward to protect the air, the underground, and the water space.

Markets & Markets in their report 'Anti-drone market—Global forecast to 2028', forecast the anti-drone market to reach from £1.3 billion in 2023 to £4.2 billion by 2028.

On 4th of December 2023 the Pentagon Administration Chief highlights the counter-drone shortage, saying 'Both Israel and Ukraine are facing an operational crises'. He requested more funds for supporting them but also for own country, emphasising how vital the counter drones are and that their production should 'surge through the roof'. The US Army has formed a Joint Counter Small Unmanned Aircraft System Office (JCO) and its director explained that 'there is no silver bullet' but the 'electronic warfare will likely play a role in fending off swarms' (JCO Director Maj. Gen. Sean Gainey).

Counter drones employ a variety of technologies in an over-arching attempt to detect, track, classify, identify, mitigate drone threats. The first step is to determine that it is a drone (not a bird etc.) with micro-doppler radar, after that a decision has to be made whether the drone is friendly or malicious. Regulators in many countries, including the UK, have implemented or are planning to impose that the drones are registered and also have remote identification (RID). This will contribute to the identification of the drones, their operators and their behaviour and help the counter drone to take a rapid decision about the next step. In case of a rogue drone, the mitigation measures can be: signal jammers, GPF spoofers, geo fencing, nets launched by a cannon or other drone, lasers, high power microwave (HPM) devices, cyber hacking. These technologies could be used together for increasing the success rate.

Shield UAS Solutions is presenting the symbiotic integration of radar, lidar and acoustic sensors with AI algorithms in order to offer a layer security and enhance the threat detection precision in a holistic way, continuously evolving way.

Radar has been proven to be a very effective over the years, but nowadays, given the size of the agility of smaller drones, it is more difficult to distinguish them from the background clutter. Here the AI offers digital processing algorithms for processing traditional radar signals, and combined with electro-optical sensors and higher frequency radar systems it improves accuracy, especially where detection is hindered by urban infrastructure.

Companies like Robin Radar Technologies are specialised in detecting small drones and birds. Lidar (Light Detection and Ranging) is based on emitting laser pulses and measuring the time taken by these signals to bounce back from the objects they reach. This is a very efficient way to distinguish between drones and other objects and create accurate 3D maps of intricate environments.

Acoustic sensors are based on capturing sound waves produced by drones in flight. ML algorithms compare various drone acoustic patterns data (frequency, pitch and duration of the sound) associated with known drone models. An example is detecting drones flying around prisons with variable weights, the number of the propellers rotations indicating heavier when getting closer to the prison, lighter when flying away, meaning potential payload dropped at the prison (Chris Chruch, from Interpol, at DroneSec GDSN, August 2021).

The acoustic drone detection has many advantages, such as generating minimal false positives, being effective in various environments, integration with other detection technologies etc.

The main challenges of combining AI with the traditional technologies are related to data privacy, compliance with regulations and environmental interference. A holistic and adaptive approach is mandatory for keeping the airspace safe and secure. As they are a relatively new technology, not only the quantity but also the quality of the counter drones needs increasing, and that is achievable by combining multiple technologies.

Khawaja and his colleagues present in their report an overview of various radar systems, one of them being the Joint Communication-radar (JCR) system, comparing them with other systems that do not involve radar for in their missions to detect, track and classify drones threats. signal jammers, GPF spoofers, geo fencing, nets launched by a cannon or other drone, lasers, high power microwave (HPM) devices, cyber hacking. These technologies could be used together for increasing the success rate (Khawaja et al., 2022).

The collaboration between various manufacturers by forming coalitions, complementing each other's technologies and skills help improving the quality and complexity of the counter drone technologies. An example is DroneShield (with DroneGun Tactical, DroneGUN MK3 and 4, DroneCannon, RFPatrol, DroneSentry-X), DroneSec (Intel platform, Stolen drones DB, certifications) and Dedrone (use of AI and ML, DedroneTactical, Radar, PTZ cameras, Dedrone Defender, Dedrone Tracker etc.). To all these technologies, one of the market leaders Robin Radar

Systems, highlights that some of the malicious drones operate in adverse conditions, so the counter-drones technologies must become more robust and withstand the elements, as in the case of protecting oil rigs etc.

And what can be more secure than a counter-drone? A team of counter drones that work together, applying multiple platforms for detecting, mitigating, and controlling mini drones according to Castrillo & Co., in their paper about teams of counter drones against cooperative teams of drones (Castrillo et al., 2022).

Not only the air but also the underwater needs to be protected. There are many technologies in the ocean and there is need for more sensors and counter drones to protect the assets under the water (submarines, data cables, oil rigs and pipes, transportation etc.) An example of such a project is the 'Silent Swarm': US Navy seeking electromagnetic spectrum tech crossed with unmanned organised by the US Army in collaboration with Australia and UK. The Navy is testing early-stage tech with capabilities relating to distributed electromagnetic attack, deception, and digital payload delivery.

DARPA is using laser to teach swarms of drones to swim autonomously like fish, to create 3D laser maps of environments like whole ships, buildings, entire urban environments, to train swarms to navigate by using ML. This program is called Offensive Swarm Enabled Tactics (OFFSET). The swarm takes 3 mil pictures per second, creating a live 3D map, the pilot with the VR googles follows the map and sends commands back to the swarm, so the drones can navigate with a combination of ML and human input.

Another technology that could contribute to the improvement of the counter drone systems is the laser. Although it has been used in the past as chemical laser, not very successfully then, the laser weapons are making a come-back now, as high energy lasers (HEL), especially as part of defence against enemy drones by integrating it into the counter drones multi-layered systems. For example, in the war, if the attacker uses a customised commercial drone that cost $2K, the defence could use very efficient missiles against it, but they could cost from $100K to $3 million each. Laser based counter drones are more cost effective and efficient too. US, UK, Russia, China, Germany, France, Israel have been testing laser weapons systems in the air and under water.

High energy lasers (HEL) produced by companies as Raytheon protect against threats posed by rockets, artillery, mortars (RAM) or low, slow, small (LSS) targets like drones swarms of drones, by melting them in seconds with their modular power up to 50 kW. At sea the laser compact laser weapon systems (CLaWS) that has been in use by US with undisclosed power, supposedly up into megawatts, it has been used since 2014.

Drones raise ethical issues too as with great power comes great responsibility. What happens when autonomous drones driven by AI target and kill humans? How can this be countered? What kind of counter drones systems are needed in that case? An United Nations Security Council report on military operations in Libya show the Turkish drone STM Kargu-2 using AI to detect and destroy human targets based on information in the database, without any commands being needed. This is significant because it is the first time that it is recorded that drones kill human beings based on artificial intelligence.

In 2019 a total of 356 Kargu-2 drones have been added to the Turkish Armed Forces inventory. The question is if machines driven by artificial intelligence should be allowed to kill humans? Or we better go to the root of the issue and ask if the military should be allowed to build those machines at all and how could this be controlled? The 'EU AI Act: first regulation on artificial intelligence' does not seem to mention it under the 'High Risk', or even mention it at all. There is only anecdotal evidence that Turkey has sold the STM Kargu-2 to three other 'clients' who remain unnamed. Hopefully this is only fabricated news, but if there is any truth in this story, then sadly more humans will die by the power of the killer drones.

Another powerful gun, from a different category, is has been developed in Japan. In October 2023 the Japanese Acquisition, Technology & Logistics Agency (ATLA) announced the successful testing of firing a ship-board railgun, this being the first in the world of its kind. This gun will counter and protect the Japanese vessels from air-threats and surface-threats posed by anything up to Mach5 and higher hypersonic missiles.

Usually drones and some counter drone operations are limited by how much their batteries hold but thanks to lasers, the drones can charge on their way, never being necessary to return to a charging station because of power beaming. This technology has been tested since 2017 and it works on submarine drones via an optical cable, but also evolved to open air power transfers that have been successfully tested on drones. This is revolutionary as power beaming is portable, silent, invisible and facilitate power transfers to less accessible areas. A new project of the Air Force Research Laboratory called the Space Solar Power Incremental Demonstrations and Research Project (SSPIDR), aims to use satellites to collect energy from the sun and beam that energy down directly to receivers on Earth, eliminating the need for power infrastructure. At the moment it is not known what happens if this power beam hits the wrong spot on Earth.

In his article David Axe describes how Ukrainian military have developed a radio system that not only jams the Russian drone signals, but also the Russian counter drones. This radio-jammer is very effective, for as long as the Russians don't develop a radio-jammer jammer ('Ukraine Has Figured Out A New Way of Winning Battles: Start By Grounding Russia's Drones', Forbes, December 2023).

These are only a few examples of current drones and counter drone technologies. Although there are some high performance, sophisticated counter drones on the market and new ones are emerging, they are not enough to cover the high demand and also they are not well regulated. The counter drone stakeholders must be a step ahead of the malicious actors. One of the challenges is that the same technologies can fall in the wrong hands and rogue operators are becoming more sophisticated. This issue needs to be addressed by regulating who has access over what technologies.

These regulations must be current and effective, without leaving open doors for ambiguities, as in the case of the Guideline and its two annexes published by the Missile Technology Control Regime (MTCR). While this is an agreement between a group of countries trying to control the exporting/importing drones and drone parts, they are still able to bypass the agreement by classifying some of the drones as 'missiles'. There is a huge need for national and international regulations, for both times of peace and times of war. 'Wait and see' is not the right approach.

2.5 *Remote Identification (RID)*

To increase the safety and security of the airspace, the implementation of the Remote ID came as a response to the increasing number of drones in the sky. Remote ID allows the authorities, law enforcement, and other authorised organisations or individuals to discover, identify and track UAS while being in flight.

'Remote ID is the ability of a UAS to communicate identification and location information during flight. The requirement for Remote ID already exists in UK Regulation (EU) 2019/945 and 2019/947 but is yet to be implemented. The Call for Input described the important role of Remote ID in improving the ability to identify individuals misusing UAS and in enabling enforcement. In addition, it discussed how Remote ID could be implemented in practice and potential changes to the regulation.' CAA Remote ID definition, Published by the Civil Aviation Authority, 2023.

From a counter drone perspective, the first step in the identification of a flying object is to determine whether it is a drone (not a bird etc.). The next step is to understand if the drone is friendly or malicious. This can be done with the help of a Remote ID system that the counter drone is able to communicate with and establish the identity of the drone and its behaviour (Belwafi et al., 2022).

The need for this type of identification has determined regulators in many countries, including the UK, to implement a system that requires the drones to be registered and also to have a type of remote identification, similar to a car number plate.

The main features of Remote ID are: authentication and authorisation; integration with the Traffic Management systems and air traffic controllers; Identification information (serial number, operator location, drone's GPS coordinates; transmitting broadcast or networked data; compliance with specific standards; complying with privacy requirements by only publishing the necessary details, without compromising the pilot's privacy; real-time monitoring; establishing No-Fly Zones (if integrated with geofencing technologies); automated alerts and notifications; public awareness.

As it hasn't reached its maturity stage, the Remote ID has some disadvantages too, but their advantages are more significant (Tables 1–4).

Effective identification combined with digital forensics technics are necessary for the counter-drones to identify the authorised and unauthorised drones, and to collect, analyse and store digital evidence related to drones incidents. The use of Remote Identification (RID) technologies will address these challenges by providing a regulatory measure.

This research is looking into how the remote identification will aid the counter drones systems to collect, store and preserve the chain of custody of digital evidence.

Table 1 Advantages of remote ID for drones, L.N. Ivascu

Advantages of Remote ID for Drones	
Enhanced Safety	Remote ID enables authorities and other airspace users to identify and track drones; enhancing safety and security. It allows for the identification of rogue or unauthorized drones, helping mitigate potential security threats and incidents
Air Traffic Management	Remote ID facilitates the integration of drones into existing airspace systems, enabling better coordination with manned aircraft. This integration can lead to more efficient air traffic management, reducing the risk of collisions and improving overall airspace safety
Accountability and Enforcement	With Remote ID, it becomes easier to hold drone operators accountable for their actions. In case of any violations or incidents, authorities can identify the responsible party through the unique identifier transmitted by the drone
Public Acceptance	Implementing Remote ID can contribute to public acceptance of drones by assuaging concerns related to privacy, security, and accountability. By providing transparency and traceability, it can help build trust between drone operators and the general public

Table 2 Disadvantages of remote ID for drones, L.N. Ivascu

Disadvantages of Remote ID for Drones	
Cost and Complexity	Implementing Remote ID systems may involve additional costs for drone manufacturers, operators, and regulatory bodies. The hardware, software, and infrastructure required for remote identification can increase the overall expense and complexity of operating drones
Privacy Concerns	Remote ID raises concerns about the potential invasion of privacy. The constant tracking and identification of drones could be seen as a form of surveillance, leading to apprehension among individuals who fear their privacy may be compromised
Technical Challenges	Developing and implementing an effective Remote ID system poses technical challenges. Ensuring accurate identification, reliable transmission and interoperability across different drone platforms and technologies can be complex
Impact on Hobbyists	Remote ID requirements may affect recreational drone users and hobbyists. The additional registration, identification and equipment requirements could potentially deter some individuals from engaging in drone-related hobbies

- **International context**

Some countries have decided to ban drones including: Algeria, Brunei, Barbados, Cuba, Cote d'Ivorie, Iraq, Kyrgyzstan, Madagascar, Morocco, North Korea, Nicaragua, Senegal, and Syria (Interpol, Framework for Responding to a Drone Incident, January 2020).

Table 3 Scenario where a rogue drone is investigated, following the drone digital forensics steps—Part 1/2, L.N. Ivascu

Step	Event	Tools	Analysis
Detection Phase	An unauthorized drone is detected flying near a critical infrastructure facility. The facility's security system triggers an alert, and authorities decide to investigate the incident	Remote ID technology is utilized to identify the rogue drone and gather information about its operator	
Incident Response	Security personnel initiate an incident response, deploying protocols to assess the threat and mitigate risks	Digital forensics experts are called in to investigate the rogue drone incident, preserving the scene and ensuring no tampering Occurs	
Isolation and capture	Security personnel work to isolate the rogue drone and capture it safely	Forensic experts provide guidance on preserving electronic evidence during the capture process, ensuring the drone's data and hardware integrity are maintained	
Extraction of Flight Logs	The captured drone's flight logs are extracted using specialized tools	UFED Drone, DJI Assistant, Drone escape Forensics	The fight logs reveal the drone's flight path, altitude charges, and potential waypoints
Imaging the Drone's Storage	A forensic investigator creates a forensic image of the drone's storage to preserve the Original data	dd or dcfldd (Command-Line Tools): Used for creating a bit-for-bit forensic image of the drone's storage	
Memory Analysis	The investigator conducts memory analysis to identify active processes and gather volatile data	Volatility Framework: Used to analyze the memory dump for running processes, open connections, and any volatile artifacts	
File Carving	File carving is performed to recover deleted and hidden files from the drone's storage	Scalpel or Foremast: Used for carving files from unallocated space or fragmented storage	

(continued)

Table 3 (continued)

Step	Event	Tools	Analysis
Analysis of Digital Media	Images and videos captured by the drone are analyzed	PhotoRec, ExifTool, FIVE	Investigators examine the media for visuals of the surrounding area, potential targets, or any suspicious activities
Communication Log Analysis	Communication logs between the rogue drone and its ground control station are analyzed	Wireshark, RFcat	Investigators look for unauthorized commands, communication patterns, and any attempts to interfere with the drone's control
Sensor Data Examination	Sensor data from the drone's onboard sensors is analyzed	Flight controller logs, sensor-specific analysis tools	Investigators assess the drone's movements, reactions to external factors, and any anomalies in sensor readings

Other countries and international organisations have made a collaborative effort and have produced various regulations for their drones and counter-drones environments, the Remote ID being on many of their agendas, such as: US, EU, India, Switzerland, Canada, New Zealand, Japan.

The 'Standard Specification for Remote ID and Tracking' published by ASTM F341119 is the standard that proposed the transmission of the Remote ID data via two channels: IP-layer (Network RID) and MAC-layer (Broadcast RID).

The ASD-STAN has published a Remote ID standard for Europe PREN 4709-002, referring to the EASA (EU)2019/945 and EASA (EU)2019/947, updated in 2022 (EASA Counter Drones (C-UAS) Action Plan, n.d.).

In March 2023 ASD-STAN has published three out of four prepared standards from the series prEN 4709, one of them being, prEN 4709-002 P1 Corrigendum 1, referring to Direct Remote Identification.

- United States

In the US the FAA have completed their consultation period already and after receiving more than 53 k responses, the Remote ID will be implemented from 16th of March 2024 (postponed from 16th of September 2023) There will be three options to comply with the Remote ID requirements: via a built in device, via a broadcast module or by flying in special designated areas where the FAA doesn't require identification. The US give the manufacturers two options to broadcast: the sessionID

Table 4 Scenario where a rogue drone is investigated, following the drone digital forensics steps—Part2/2, L.N. Ivascu

Step	Event	Tools	Analysis
Firmware Analysis	The drone's firmware is extracted and analyzed	Binwalk, IDA Pro	Investigators examine the firmware for any unauthorized modifications, backdoors, or alterations to default settings
Battery and Power System Analysis	Data related to the drone's battery status and power system are analyzed	Battery management system tools multimeters	Investigators assess whether power-related issues contributed to the drone's behavior
Network Activity Investigation	Network logs and communication patterns with any remote connections are analyzed	Wireshark, Network IDS/IPS	Investigators look for any network connections, data exchanges, or attempts to connect to unauthorized networks
Timeline Analysis	A timeline analysis is conducted to establish a chronological sequence of events related to the drone's activities	Autopsy or The Sleuth Kit: Used for timeline analysis, incorporating information from various artifacts	
Keyword Search	Investigators perform keyword searches on the extracted data to identify relevant information	Autopsy or grep (Command-Line Tool): Used for searching keywords within the file system	
Analysis of Metadata	Metadata analysis is conducted to understand the origin and history of files	ExifTool: Used to extract metadata from images and other files	
Forensic Reporting	The forensic investigator compiles findings and prepares a detailed forensic report for the court	Executive Summary: Summarizes key findings	

(continued)

Table 4 (continued)

Step	Event	Tools	Analysis
Components of the Report	Executive Summary: Summarizes key findings	Technical Details: Detailed technical analysis, including flight logs, firmware analysis, network activity, and extracted artifacts; visual evidence form images and videos; battery assessment; communication logs; sensor data findings	
Conclusions and Recommendations	Provides conclusions based on the evidence and suggests further actions	Appendix: Includes detailed logs, screenshots, and supporting documentation	
Presentation to the Court	The investigator presents the forensic findings in court, providing expert testimony about the rogue drone, explaining their methodology and making sure that they preserve the authenticity and reliability of the digital forensic evidence	Presentation Software: PowerPoint or similar tools for presenting forensic evidence, digital forensic templates	

(that gets renewed with every take off) or the unique identifier of the drone. The transmission will be every second, but it is not specified what technology will be used.

The data transmitted will be: identification, location of the drone, velocity, location of the control station, time mark, emergency status only for the built in device), and the serial number of the attached module, where is the case.

- European Union

In the European Union the central point of reference for the implementation of the remote ID is EASA. From 1st of January 2024 all drones with class marks and flying in the open category and all drones in the specific category, will be required to operate with an activated Remote ID. The data transmitted will be: the operator registration number, the location of the operator, the location of the drone, the hight above the take-off point, the route course measured from the north, the velocity, the classes C0 to C4 are in the ‘Open’ category. This category might vary slightly per EU state, for example the minimum age of the operator.

The drones that are not already equipped with Remote ID will have a Remote ID module attached to them, according to the EU Declaration of Conformity list. No requirements are mentioned for data transmission intervals or technology used.

As UK has left the EU, this Remote ID rule does not apply in the UK, but the UK residents who travel to Europe with their drones must be aware and comply with the EU regulation.

- UK Strategies

In the UK, to improve the safety of the airspace, the CAA is planning to implement the remote identification (RID) for drones too. On request of the Secretary of State, the CAA and the Department of Transport (DoT) have developed the Airspace Modernisation Strategy for 2023–2040. The most important part of the strategy is 'improving the UK's high levels of aviation safety', one of the drivers being to 'facilitate defence and security objectives'. Some of the benefits of the modernisation of the airspace would be to 'accommodate new technology and new or rapidly developing types of aerial vehicles, such as drones providing new services to businesses and the public, alongside medical or search and rescue flights', 'while maintaining and, where possible, improving the UK's high levels of aviation safety.'

On 9th of August 2023 the CAA invited the drone community to send their feedback related to the Remote ID in UK. Based on the 2629 responses from the Call for Input received by the 7th of September 2023 and with the input from the Home Office, Department of Transport and the Police, on 23rd of November 2023 the CAA has launched a consultation on a proposal to develop regulations related to the introduction of the Remote ID in UK.

Kevin Woolsey, co-head of Remotely Piloted Aircraft Systems at the UK Civil Aviation Authority, said:

> We want to make sure that the use of drones continues to safely expand and that regardless of whether people are flying for fun or commercially they understand the rules and regulations.

After the quantitative and qualitative analysis of the feedback, the CAA concluded that there are some opportunities for improvement.

One of the most discussed changes (it appears 79 times in the proposal) is the implementation of the Remote ID, which together with geo-fencing and geo-awareness would enable the safe and secure airspace. This opportunity for change was received mixed views, 52.4% of the participants saying 'definitely no' to the implementation of the Remote ID.

Some of the proposed changes are related to improving drone safety and security by introducing a class-marking for drones; simplifying regulations by removing confusion about the 'open' and also the 'toy' terms; increasing education and understanding of regulations; supporting the drone sector by giving manufacturers and UAS operators more time to adopt the new regulations.

At the time of writing this report, the UK CCA is in consultation until the 10th of January 2024.

The stakeholders involved are most likely organisations involved in the projects mentioned in the Airspace Modernisation Strategy 2023–2040 and also the future emerging projects involving the UK airspace, these projects being SkyWay, Atypical Airspace BVLOS (Beyond Visual Line of Sight) Solution (AABS), InDepth, EagleX. All these projects will massively benefit from the implementation of the Remote ID in

UK. Because the consultation is still ongoing, the first question that arises is whether the implementation of the Remote ID will get green light or not.

2.6 Remote ID Framework

In the situation that the decision is taken to carry on with the implementation of the Remote ID, a framework will be developed and published. That will involve establishing guidelines, procedures, and standards in order to ensure the successful deployment and use of the Remote ID for UAVs. These would be the key points to consider when writing the framework:

- Ensure compliance with the existing regulations related to UAV and airspace management.
- Clearly define the scope and objectives of the remote ID framework.
- Establish the standards for the technologies used in Remote ID systems.
- Adress privacy and security concerns of Remote ID data.
- Define the data governance, how the Remote ID data will be collected, managed, and shared.
- Ensure interoperability with other systems involved in the process.
- Implement timeline for implementing Remote ID, using a phased approach.
- Engage relevant stakeholders.
- Inform the public and the drone operators about Remote ID.
- Define enforcement mechanisms and penalties in case of non-compliance.
- Implement a testing and certification process for the remote ID systems.
- Establish how the framework will be periodically updated and improved.

To reach the above objectives, CAA will need to consider consulting and aligning with other aviation authorities existing and upcoming frameworks and define the roles of the stakeholders involved. The most relevant regulations are the ones provided by the US (FAA) and Europe (EASA) as mentioned here above.

CAA will also need to define what types of drones are covered by the framework, what geographic areas are covered, the technical details of the transmission frequencies, data formats, communication protocols, what types of devices are permitted to have remote ID, how to deal with emerging technologies etc.

The CAA will need to ensure privacy and security by defining guidelines for access to the Remote ID information and data storage, what type of encryption(s) to use and how to protect the operator data, what are the retention policies, what data is going to be collected, and if and when the data to be shared with authorities.

Because of its involving with many other systems, CAA will need to make sure that the interoperability between them exists, that manufacturers collaborate and follow common standards.

CAA will also need to identify the stakeholders and keep them involved via efficient communication channels, organise trainings and information materials for them and for the public.

Most of the people will comply, but for the ones who don't there should be defined procedure of how to report and investigate incidents and what consequences there will be for who will not comply.

To ensure the safety but also the quality of the remote ID systems, CAA should outline how the testing and maintenance, but also the certification of the Remote ID systems will be done.

As this is a relatively new domain, with new developments continuously evolving, the CAA will need to define how to deal with the stakeholders' feedback and what would be the mechanisms to keep the framework updated.

Some of the questions that CAA will need to answer are:

- What information will the Remote ID bring in?
- How will CAA accommodate evolving technology standards?
- What measures will be in place to prevent unauthorised access to the Remote ID systems?
- How will the challenges related to the compatibility with other systems be addressed?
- What is the timeline and what are the critical milestones for the implementation?
- How are the penalties for non-compliance defined and who will enforce them?
- How will the framework be adapting to the evolving technologies?
- What are the procedures in case of a large scale emergency, or even war?

The consultation is on until the 7th of January 2024, then the CAA will form a final conclusion, then it will present it to the Ministry of Transport and submitted to the Government for being approved, and finally implemented. A gradual implementation would be recommended so the stakeholders have sufficient time to prepare.

In the meantime, on 1st of January 2024 the eyes will be on the EU as the Remote ID rules will start to apply. It will be very interesting to observe what goes well and what improvement points there are across Europe, so the FAA, CAA and other international organisations can benefit from those learning points.

2.7 Drones Digital Forensics

Since the beginnings of the Police, not much has changed regarding the chain of events in case of a (potential) crime has happened (Fig. 2).

This was true until the drones have appeared and with them the drone criminality in its various shapes and forms. During the Global Drone Security Network (GDSN) focusing on Cyber-UAV security, Drone Threat Intelligence, Counter-UAS, Christopher Church, a senior forensics specialist from Interpol was highlighting the way the drones are changing the way the police is handling a crime. In the past, the criminals could be physically constrained and prosecuted. Nowadays the nefarious actors could be in another country, controlling the drone remotely and maybe with the help of a 'mule'. Also, the rogue drones could be controlled by personal mobile phones that are not easy to seize and access. There are not enough skilled police

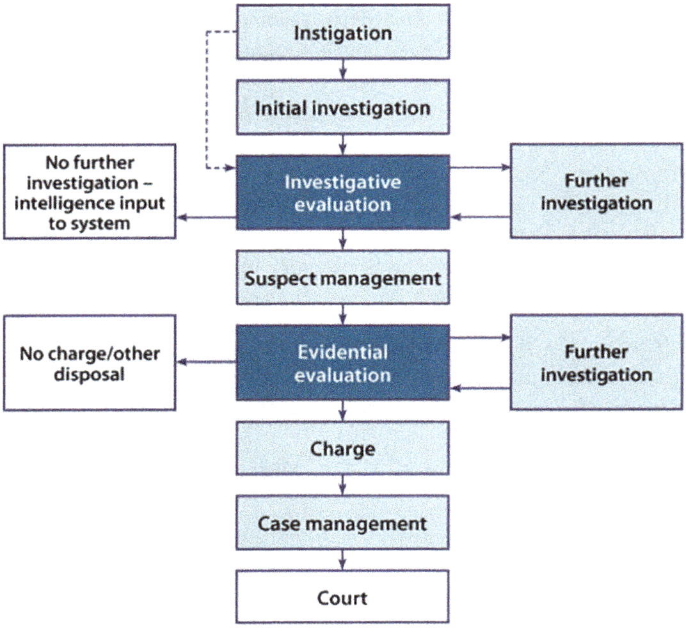

Fig. 2 Traditional digital forensics (*Source* Interpol)

forces, insufficient funds for training and often if they get trained, they move on to join the private sector. Drones are increasingly used to smuggle contraband items into prisons. In US alone there were 2500 drones seized over a period of 18 months. and the majority of these drones were not further digitally investigate further than the wet forensics (finger prints etc.) due to the drones digital forensics skills shortage and the lack of appropriate regulations. Often the reaction plans that the police forces formulate are not tested or they don't work, as in the case of the drone incident that caused Gatwick Airport to shut down around the Christmas time in 2018. In addition to that, worldwide more than 100 drone incidents involving airports were reported in 2021 and the first quarter of 2022, their motivations being either pure negligence, reckless or careless individuals, protesters/activists or criminal/terrorist intentions.

In light of these challenges the Interpol has organised the Project Courageous, that has been taking place between 2021 and 2023, funded by the European Union Internal Security Fund Police and aiming to 'develop a standardized test methodology for detecting, tracking and identifying illicit drones' (Interpol). The participants at this project are law enforcement and research and technology partners from eleven European countries. The exercise took place at a Norway national airport, while 2025 aircraft took off and landed over a three-day period of time and the conclusions were published in the 'Interpol Drone Countermeasure Exercise Report' (INTERPOL DRONE COUNTERMEASURE EXERCISE REPORT A Report Jointly Produced by INTERPOL and Norwegian Police, 2022). It is a comprehensive (72 pages) report, based on a complex exercise, unique in its way, given the amplitude of the whole process in such a significant environment.

Another valuable document produced by the Interpol is the 'Framework for Responding to a Drone Incident, For First Responders and Digital Forensics Practitioners', published in January 2020, setting the foundations for the drone digital forensics.

> Drones digital forensics refers to the process of acquiring, analysing, and interpreting digital evidence from unmanned aerial vehicles (UAVs), commonly known as drones, in the context of forensic investigations. It involves applying forensic techniques and methodologies to extract and analyse data from various components of a drone, including its hardware, software, communication systems, and storage devices.

The typical chain of custody in drones digital forensics is: identification, collection, packaging, labelling, documentation, transportation, receiving in the laboratory, analysis, documentation of analysis, storage, court presentation, return or disposal. It is paramount to maintain the integrity of the chain so that the digital evidence is authentic, reliable, credible, and admissible in legal proceedings.

The data could be flight logs collected form the on board flight controllers and GPS modules (altitude, GPS coordinates, speed, motor outputs, battery status and other sensors data during the flight); digital media from onboard sensors and cameras; communication logs from the transceiver modules, communication protocols and ground control stations (telemetry, records of communications, status updates etc.), sensors data from onboard devices as magnetometers, accelerometers, gyroscopes and environmental sensors), data from firmware as the internal storage (algorithms, configurations, flight modes), ground control station data (telemetry, planning details, user inputs etc.), network activity between drone and external networks (IP addresses, communication protocols, network logs),battery information (charging status, voltage levels, flight duration etc.). Example of the batteries used by a group of prison smugglers, identified by using one battery to many drones.

The data could be collected by physical examination of the drone, analysis of the ground control station, remote data extraction, wireless network monitoring, forensic imaging, in some cases using blockchain technology and other collaboration platforms for a comprehensive approach (Renduchintala et al., 2018).

The challenges faced by the drone data collection are the encryption and security measures, data loss risk by damage, proprietary systems and legal and ethical considerations related to compliance with privacy and aviation regulations.

Some of the digital forensic tools are FTK Imager, EnCase, dd, Sleuth KitAutopsy, X-Ways Forensics, Forensic Toolkit, AccessData Forensics Toolkit (FTK), Volatility, Paladin Forensic Suite, Cellebrite UFED (for drones), BlackBox Log Viewer, RFcat, Wireshark, SDR, FIVE, Binwalk, IDA Pro, Hex editors etc. Some drones manufacturers like DJI offer tools to access the drone configuration data and perform diagnostics (DJI Assistant) (Yousef et al., 2020).

The main drone digital forensics methodologies used are: ACPO (Association of Chief Police Officers) Guidelines; ISO/IEC 27037:2012; NIST (National Institute of Standards and Technology) Digital Forensics Framework; ACE-V (Analysis,

Comparison, Evaluation, and Verification); OSCO (Objective, Scope, Context, and Output); Hacking Incident Response Plan (HIRP); Digital Forensic Research Workshop (DFRWS) Methodology; Computer and Enterprise Investigations Guide (CEIG) (Thornton & Bagheri Zadeh, 2022).

In addition to the Interpol framework mentioned earlier, other frameworks to be followed by the drone forensics investigators are:

- Digital investigations frameworks: SANS Investigative Forensic Toolkit (SIFT); Open Computer Forensics Architecture (OCFA); Digital Forensic Research Workshop (DFRWS) Framework; National Institute of Standards and Technology (NIST) Cybersecurity Framework.
- Incident response frameworks: NIST Computer Security Incident Handling Guide (SP 800-61 Rev. 2); FIRST (Forum of Incident Response and Security Teams) Framework for CSIRTs;
- Network forensics frameworks: Security Onion; Network Forensic Analysis Tool (NFAT).
- Mobile Forensics Frameworks: Mobile Forensics Framework (MobSF); XRY (Mobile Data Forensic Phone Recovery & Extraction).

Some of the challenges (Azhar, 2019) that the law enforcement organisations are facing are related to the fact that technologies get developed faster than the legislation being updated. For example, there are drone taxies, drone cargo planes, flying cars etc. that are not currently regulated, and it doesn't seem to be a coordinated discussion around them. The regulators talk with the industry and later they expect the law enforcement to maintain the security of the airspace, without having enough resources. Criminal network mindset and technology adapt quicker than the police forces do. Drones and their components are cheap to purchase and there are numerous tutorials on internet about how to build rogue drones for various types of attacks. Autonomous drones and frequency hopping make the police work even more complicated.

To close the law enforcement shortcomings gap, Interpol created a Drone Expert Group with participants form the 194 member countries and a Drones Forensics Lab (DFL, in Singapore). They are open to collaboration with all members and share knowledge and skills across the borders.

At the time of writing this paper, there are 2984 Kali Linux pen tools, 171 being related to forensics and only 4 specifically to drones digital forensics. Although the 'traditional' digital forensic tools are good enough in some situations, there is a need to adapt to the continuously developing of malicious technologies and create dedicated drones/counter-drones digital forensics tools.

Drone digital forensics, also a relatively new and rapidly evolving field that deals with collecting, analysing, interpreting, preserving, and reporting digital evidence from drones, is presenting a considerable range of research gaps too, some of them being mentioned in this paper.

Drone network forensics such as communication protocols, network traffic analysis, identification of digital footprints, drones network infrastructure analysis, visualisation and reporting, network traffic analysis are crucial in forensic investigations.

Various communication technologies have been explored and need more research, such as Wi-Fi (IEEE 802.11a/b/g/n/ac); cellular networks (GSM, CDMA, LTE to analyse IMEI, IMSI etc.); dedicated RF; C2 protocols (MAVLink, DJI GO etc.); encrypted communication mechanisms etc.

Forensic analysis of continuously emerging drone technologies could bring new challenges and opportunities by involving cutting edge drone technologies. There is a shortage of drone forensic labs and tools to perform such analysis. Although drones digital forensics are basically similar to the traditional digital forensics, there are some differences that need to be addressed. At the time of writing this report there are 2984 Kali Linux tools, but only a few are specifically designed for drones (Crozono, Missionplanner, SkyJack, Snoopy-ng).

Image and video analysis techniques need more research related to the crucial role of the forensic evidence captured by drones, including automated identification of people, objects or locations, also detection of tampering with visual or audio evidence captured by drones. Photogrammetry, 3D mapping.

Specialised training and education in drones digital forensics is needed because of the increased presence of drones in various fields, such as disaster response, law enforcement, infrastructure inspection etc. to enable forensic investigators to conduct correctly drone related activities.

Standardized methodologies and best practices are needed for developing standardised methodologies specific to the digital forensics industry. Widely accepted frameworks or guidelines for collecting, analysing, and interpreting digital evidence from drones and counter drones are currently missing. Nefarious actors are active across borders, so that's how digital forensics investigators should collaborate too (Alotaibi, Al-Dhaqm, & Al-Otaibi, 2022; Alotaibi, Al-Dhaqm, Al-Otaibi, et al., 2022; Ojo et al., 2022).

Privacy and legal implications represent another research gap due to the ability of drones to collect data from various environments, often not being clear what the legal implications are related to the privacy laws, data collection and data sharing. Legal frameworks, guidelines and ethical considerations need to be in place and be current in order to keep the pace with the drones development. Policy and governance has been another significant field explored by academics, trying to ensure that the technology is used in a responsible way and to raise awareness about the legal and ethical implications of counter-drone technologies, such as balancing security with individual rights, privacy, airspace regulations, collateral damage etc. (Svaigen et al., 2022).

Although the research of drones is expanding rapidly, counter-drones are still a young technology and there are still gaps in the knowledge that could be explored.

More research is needed regarding the impact of the countermeasures on various types of drones, their impact on the environment and the infrastructure, and also on the safety of the surrounding human population. The public perception of counter-drones needs improving, so the public understands that 'counter-drones' are not 'against drones', but 'pro safety'.

2.8 *Rogue Drone Scenarios*

Drones are used for various malicious purposes. Although most of the drone digital forensics process steps are similar to the traditional digital forensics, the Remote ID plays a crucial role in the beginning of the scenario, when the drone gets detected and needs to be identified. Without the Remote ID it would be more complicate or sometimes impossible or not worth finding the operator/owner of the drone.

This following scenario presents the steps of a typical process of detecting and investigating a rogue drone incident, using digital forensics and Remote ID, adapted from the traditional digital forensics chain of actions.

3 Potential Future Research Opportunities

How would remote identification technologies aid the counter-drones digital forensics in general, and in preparation for the implementation of the UK's Airspace Modernisation Strategy 2023–2040 in particular? The question is answered partially because Remote ID in UK is not implemented yet and the results from other countries are not yet conclusive. The question remains open, as an opportunity for future researchers to investigate further and find the answers.

The research refers to drones and counter drones, their types, their components, their communicating ways, their activities, in relation to digital forensics assisted by Remote ID. Within the scope of the research there are also international and UK regulations related to drones, counter drones, digital forensics and Remote ID.

The topic of this research is very interesting and complex, as it combines a wide range of areas of interest. Because it's a novel, vast evolving subject, the academic resources are quickly 'old news'. By the time the academic research is done methodically, the engineers and hobbyists, often in the war zone, discover, test and adapt new technologies that, although ingenious, never appear in a scientific paper, being executed without a risk management, lab tools and materials, designated test environment, peer reviews etc.

The developments in drones and counter drones are often unveiled via the mainstream media, because it all happens too fast. The risk of using this type of sources is that news could be fake or inaccurate. Is it advisable to check the background of the writers of these articles, as they are often contributions to newspapers and magazines made by the same subject matter experts and scientists who also write good quality academic papers.

Drones and counter drones are fascinating, and they offer a huge potential for researcher material. During the literature review for this paper some ideas for future research have formed, each one of them being an appealing topic if the researcher would be to continue the journey into this field, or hopefully for other researchers interested in drones and counter drones (Table 5).

Table 5 Potential future research opportunities, L.N. Ivascu

What are potential future research opportunities?	
AI and Machine Learning for Counter-Drone Technology	Investigate the use of advanced machine learning and artificial intelligence techniques for more efficient and accurate counter-drone systems. Develop AI algorithms to predict and preemptively respond to drone threats based on historical data and behavior patterns
Counter-Drone Swarm Strategies	Research swarm intelligence and coordination algorithms to enhance the effectiveness of counter-drone swarms in detecting and neutralizing hostile UAV swarms
Detection of Stealthy Counter-Drones	Explore innovative methods for detecting counter-drones designed to evade traditional detection systems, such as those with minimal radar cross-section or silent propulsion
Cybersecurity for Counter-Drone Systems	Develop robust cybersecurity measures to protect counter-drone systems from hacking, spoofing, and other cyber threats. Investigate the potential vulnerabilities in communication protocols used by counter-drone systems
Secure Communication and Networking	Research communication techniques for secure and reliable communication between counter-drones, operators, and control stations, particularly in high-density urban environments and during jamming attacks
Legal and Ethical Implications	Examine the legal and ethical issues surrounding counter-drone technology, including privacy concerns and the rights of drone operators. Investigate international regulations and standards for the use of counter-drone systems
Interference and Jamming Mitigation	Explore methods to mitigate the effects of interference and jamming on counter-drone systems, ensuring they can operate effectively even in challenging electromagnetic environments
Counter-Drone Applications	Research the practical applications of counter-drone technology beyond security, such as in agriculture (e.g., for crop monitoring) and environmental monitoring (e.g., wildlife conservation)
Energy Efficiency and Sustainability	Investigate methods for improving the energy efficiency of counter-drone systems, including the use of renewable energy sources and lightweight materials

(continued)

Table 5 (continued)

What are potential future research opportunities?	
Human-Drone Interaction	Study how humans interact with counter-drones, considering factors such as user interfaces, situational awareness, and human factors engineering
Education and Training	Develop effective training programs and simulations for operators of counter-drone systems, allowing them to respond efficiently to various threat scenarios
Standardization and Collaboration	Promote collaboration among industry, academia, and government agencies to establish common standards, best practices, and information sharing for counter-drone technology
Economic and Market Analysis	Analyze the economic impact and market dynamics of the counter-drone industry, including growth opportunities and potential market disruptions
Drone Traffic Management	Research solutions for managing the increasing volume of drones in urban airspace, including collision avoidance, traffic routing, and integration with existing air traffic control systems
Biological Counter-Drone Solutions	Investigate the potential use of biological and ecological methods for detecting and mitigating drones, such as using trained birds or insects

4 Conclusions

With the latest war events in the world, technology advances at a very high pace, while the legislation takes time to catch up, going through certain stages before it gets adopted.

This chapter presents the pivotal role that remote identification plays in improving counter drone digital forensics, as the implementation of advanced remote identification technologies aids both the detection of potential threats, but also the post-incident investigation.

This research contributes to a better understanding of the drones and counter drone landscape, necessary to protect the airspace, while balancing innovation and regulation in a symbiosis between technology developments and regulatory frameworks and highlights the importance of a holistic security approach, combining advanced counter drone technologies with a digital forensic framework, augmented by remote identification.

As we can predict that drones and counter drones are not going anywhere and their number will only increase, the question is how do we make sure that our air space is safe and secure? Remote ID seems to be part of the solution but as it's still in its infancy it is difficult to measure the results of its implementation yet.

Due to the advancements in counter drones and remote identification technologies, it is recommended to keep regulatory frameworks continuously updated, and in line with other industries regulations, at national and international level.

The researchers need to address the ongoing new vulnerabilities by keeping updated with the newest developments in the drones and counter drone environment, with the latest geo-political developments in the world, but also with the newest technologies, such as quantum computing, AI, autonomous systems etc. Now AI and ML are becoming more present in counter-UAS technologies, making them more accurate and efficient, and quantum computing implications will be very interesting to watch in the next few years. Not only the scientist will need to keep updated with the latest researches but also the security personnel should receive specialised training programs. Collaboration between Government, industry and academia is paramount to establish standards for the manufacturers and service providers to ensure an effective implementation of remote identification across various drones and counter drones platforms.

The interaction between humans and drones needs to be more researched and understood but also the interaction between machine and machine is an interesting challenge.

Perhaps just like in the present there is a numberplate on every car on the street, in the near future one or more forms of Remote Identification will be present with every object that goes in the air and maybe under water too. The future started already.

References

Alotaibi, F. M., Al-Dhaqm, A., & Al-Otaibi, Y. D. (2022). A Novel Forensic Readiness Framework Applicable to the Drone Forensics Field. *Computational Intelligence and Neuroscience, 2022,*. https://doi.org/10.1155/2022/8002963

Alotaibi, F. M., Al-Dhaqm, A., Al-Otaibi, Y. D., & Alsewari, A. A. (2022). A Comprehensive Collection and Analysis Model for the Drone Forensics Field. *Sensors, 22*(17). https://doi.org/10.3390/s22176486

Azhar, H. (2019). Challenges and Techniques in Drone Forensics. https://www.dedrone.com/resources/incidents/all

Bai, Y., Lu, L., Cheng, J., Liu, J., Chen, Y., & Yu, J. (2020). Acoustic-Based Sensing and Applications: A survey. *Computer Networks, 181*, 107447. https://doi.org/10.1016/j.comnet.2020.107447

Belwafi, K., Alkadi, R., Alameri, S. A., Hamadi, H. Al, & Shoufan, A. (2022). Unmanned Aerial Vehicles' Remote Identification: A Tutorial and Survey. In *IEEE Access* (Vol. 10, pp. 87577–87601). Institute of Electrical and Electronics Engineers Inc. https://doi.org/10.1109/ACCESS.2022.3199909

Bouafif, H., Kamoun, F., & Iqbal, F. (2020). Towards a Better Understanding of Drone Forensics: A Case Study of Parrot AR Drone 2.0. *International Journal of Digital Crime and Forensics, 12*(1), 35–57. https://doi.org/10.4018/IJDCF.2020010103

Castrillo, V. U., Manco, A., Pascarella, D., & Gigante, G. (2022). A Review of Counter-UAS Technologies for Cooperative Defensive Teams of Drones. *Drones, 6*(3), 65. https://doi.org/10.3390/drones6030065

Dani, R. R., Kubde, R. R., & Sadhu, M. A. (2021). A Comprehensive Study of Various Techniques for Hostile Drone Detection and Their Classification. In *2021 6th International Conference on Communication and Electronics Systems (ICCES)* (pp. 1538–1543). https://doi.org/10.1109/ICCES51350.2021.9489236

Durfey, N., & Sajal, S. (2022). A Comprehensive Survey: Cybersecurity Challenges and Futures of Autonomous Drones. In *2022 Intermountain Engineering, Technology and Computing (IETC)* (pp. 1–7). https://doi.org/10.1109/IETC54973.2022.9796881

EASA Counter Drones (C-UAS) Action Plan. (n.d.).

Fitwi, A. H., Nagothu, D., Chen, Y., & Blasch, E. (2019). A Distributed Agent-Based Framework for a Constellation of Drones in a Military Operation. *Winter Simulation Conference (WSC), 2019*, 2548–2559. https://doi.org/10.1109/WSC40007.2019.9004907

Hambling, D. (2015). *Swarm Troopers: How Small Drones Will Conquer the World*. Archangel Ink.

Interpol Drone Countermeasure Exercise Report. (2022). A Report Jointly Produced By INTERPOL and Norwegian Police.

Khawaja, W., Ezuma, M., Semkin, V., Erden, F., Ozdemir, O., & Guvenc, I. (2022). A Survey on Detection, Tracking, and Classification of Aerial Threats using Radars and Communications Systems.

Niyonsaba, S., Konate, K., & Soidridine, M. M. (2023). A Survey on Cybersecurity in Unmanned Aerial Vehicles: Cyberattacks, Defense Techniques and Future Research Directions. *International Journal of Computer Networks and Applications, 10*(5), 688. https://doi.org/10.22247/ijcna/2023/223417

Ojo, T., Chi, H., & Erskine, S. K. (2022). Unmanned Aerial Vehicle Forensics Investigation Performance under Different Attacks. In *Proceedings—2022 International Conference on Computational Science and Computational Intelligence, CSCI 2022* (pp. 958–964). https://doi.org/10.1109/CSCI58124.2022.00171

Parkinson, S., Ward, P., Wilson, K., & Miller, J. (2017). Cyber Threats Facing Autonomous and Connected Vehicles: Future Challenges. *IEEE Transactions on Intelligent Transportation Systems, 18*(11), 2898–2915. https://doi.org/10.1109/TITS.2017.2665968

Renduchintala, A. L. P. S., Albehadili, A., & Javaid, A. Y. (2018). Drone Forensics: Digital Flight Log Examination Framework for Micro Drones. In *Proceedings—2017 International Conference on Computational Science and Computational Intelligence, CSCI 2017* (pp. 91–96). https://doi.org/10.1109/CSCI.2017.15

Rugo, A., Ardagna, C. A., & Ioini, N. E. (2023). A Security Review in the UAVNet Era: Threats, Countermeasures, and Gap Analysis. *ACM Computing Surveys, 55*(1), 1–35. https://doi.org/10.1145/3485272

Short, F. R. (2019). *A Comprehensive Approach to Countering Unmanned Aircraft Systems And Why Current Initiatives Fall Short*. JAPCC.

Shrit, O., Martin, S., Alagha, K., & Pujolle, G. (2017). A new approach to realize drone swarm using ad-hoc network. In *2017 16th Annual Mediterranean Ad Hoc Networking Workshop (Med-Hoc-Net)* (pp. 1–5). https://doi.org/10.1109/MedHocNet.2017.8001645

Siddiqi, M. A., Iwendi, C., Jaroslava, K., & Anumbe, N. (2022). Analysis on Security-Related Concerns of Unmanned Aerial Vehicle: Attacks, Limitations, and Recommendations. *Mathematical Biosciences and Engineering, 19*(3), 2641–2670. https://doi.org/10.3934/mbe.2022121

Svaigen, A. R., Boukerche, A., Ruiz, L. B., & Loureiro, A. A. F. (2022). Is the Remote ID a Threat to the Drone's Location Privacy on the Internet of Drones? In *Proceedings of the 20th ACM International Symposium on Mobility Management and Wireless Access* (pp. 81–88). https://doi.org/10.1145/3551660.3560914

Swinney, C. J., & Woods, J. C. (2022). A Review of Security Incidents and Defence Techniques Relating to the Malicious Use of Small Unmanned Aerial Systems. *IEEE Aerospace and Electronic Systems Magazine, 37*(5), 14–28. https://doi.org/10.1109/MAES.2022.3151308

Thornton, G., & Bagheri Zadeh, P. (2022). An Investigation into Unmanned Aerial System (UAS) Forensics: Data Extraction & Analysis. *Forensic Science International: Digital Investigation, 41*, 301379. https://doi.org/10.1016/j.fsidi.2022.301379

Yousef, M., Iqbal, F., & Hussain, M. (2020). Drone Forensics: A Detailed Analysis of Emerging DJI Models. In *2020 11th International Conference on Information and Communication Systems, ICICS 2020* (pp. 66–71). https://doi.org/10.1109/ICICS49469.2020.239530

Unmanned Aerial Vehicles (UAVs): Forensic, Privacy, and Security Challenges in the Era of Drone Proliferation

Georgina Butler and Reza Montasari

Abstract The proliferation of Unmanned Aerial Vehicles (UAVs), commonly known as drones, across recreational, commercial, and military domains has led to a corresponding increase in their exploitation for malicious purposes. This trend not only poses challenges to national security by encroaching upon citizens' privacy and safety but also fundamentally alters the landscape of contemporary warfare. While drones were traditionally confined to military applications, their widespread adoption by the public is fuelled by their diverse functionalities and appealing features. Drones, available in various shapes and sizes, cater to a spectrum of uses across different sectors. Nonetheless, the allure of drones is accompanied by inherent challenges and vulnerabilities, rendering them susceptible to exploitation by malicious actors. This chapter aims to provide a critical analysis of forensic, privacy, and security concerns, alongside an examination of existing legislation and regulations governing UAV usage and abuse. In light of these issues, recommendations will be proposed to address the evolving landscape of drone misuse. Additionally, an outlook on the future of drones will be presented, with a particular focus on the advancement of autonomy and the emergence of drone swarms, thereby elucidating their potential ramifications on the nature of future warfare.

Keywords Drones · Unmanned Aerial Vehicles (UAVs) · Military · Commercial · Drone swarms · Drone forensics · Civil Aviation Authority (CAA) · Regulations · Autonomy

---

G. Butler
Norwich, UK

R. Montasari (✉)
School of Social Sciences, Department of Criminology, Sociology and Social Policy, Swansea University, Singleton Park, Swansea SA2 8PP, UK
e-mail: Reza.Montasari@Swansea.ac.uk
URL: http://www.swansea.ac.uk

H. Jahankhani et al. (eds.), *Space Governance*, Space Law and Policy,
https://doi.org/10.1007/978-3-031-62228-1_8

1 Introduction

Recent technological advancements, particularly in Artificial Intelligence (AI), have significantly bolstered the capabilities of aviation and military robotics. This surge in technological prowess has not only spurred an uptick in UAV acquisitions but has also attracted the attention of criminal elements seeking to exploit these innovations for nefarious purposes. UAVs, also referred to as Remotely Piloted Aircraft Systems (RPAS) or simply drones (Stöcker et al., 2017), have become a focal point of both legitimate and illicit activity. The regulatory history of UAVs traces back to the development of manned aviation and aircraft evolution during World War II. The establishment of the Chicago Convention in 1944 marked a pivotal moment in the formulation of international aviation standards (Stöcker et al., 2017).

Despite these early initiatives, UAV regulations remain in a nascent stage, largely reactive and subject to varying degrees of enforcement across different jurisdictions. The proliferation of drones among individuals and criminal entities is reshaping the landscape of warfare, ushering in an era characterised by heightened sophistication and unpredictability (Marcus, 2022). The autonomous capabilities of UAVs, enabling operation without direct human intervention, pose challenges for policymakers and governments, prompting a reassessment of regulatory frameworks.

Drones boast numerous technical and operational advantages, exemplified by their autonomy and versatility, yet they are also fraught with vulnerabilities such as susceptibility to malware infection and spoofing. The malicious exploitation of UAVs consists of a spectrum of activities, from hijacking for targeted assaults to the illicit transportation of contraband such as weapons or narcotics (Yaacoub et al., 2020).

In light of the foregoing discussions, this chapter aims to explore the diverse applications of UAVs, spanning both commercial and recreational spheres, while also analysing how criminal elements leverage these aircraft for personal gain. Of particular concern is the role of drones in the arsenal of terrorist organisations, which is reshaping the landscape of modern warfare. Furthermore, given that the malicious uses of drones pose significant challenges in the context of national security, this chapter also examines forensic, privacy, and security issues, with a specific focus on forensic investigation protocols and the protection of individual privacy rights. Furthermore, the paper will delve into the existing legislative and regulatory frameworks governing UAV operations, with a spotlight on the oversight provided by entities such as the Civil Aviation Authority (CAA). Similarly, it will also underscore the intricate legal landscape engendered by malicious drone activities, wherein diverse legal statutes may be applicable across different cases.

Following this analysis, the chapter offers recommendations and highlights potential shifts in research focus to address the evolving dynamics of drone technology. Moreover, it will contemplate the future trajectory of drones across recreational, commercial, and military domains.

2 Background

2.1 UAV Uses

The accessibility and adaptability of drone technology have precipitated a surge in purchases by both the general public and commercial entities (Horsman, 2016). Formerly confined to military and surveillance applications, the expanding use of UAVs underscores their diverse configurations and functionalities, tailored to myriad purposes. Commercial adoption spans a broad spectrum, encompassing search and rescue operations, disaster response, wildlife monitoring, asset protection, drone journalism, and media support (Mitchell & Chen, 2014).

Similarly law enforcement agencies (LEAs) leverage drones for area surveillance, reconnaissance, patrols, and data collection. In agricultural settings, drones assist farmers in tasks such as crop monitoring, wildlife reserve surveillance, and deterrence of poaching activities (Horsman, 2016). Notably, Amazon's Prime Air initiative envisions drone deliveries for parcels meeting specified weight and size criteria, demonstrating the potential of drones in logistics (Horsman, 2016). Furthermore, drones play a pivotal role in content creation for various purposes, including filmmaking, documentaries, promotional material, and news coverage (Mitchell & Chen, 2014).

Personal use of drones for filming content is widespread among the public, with the majority of these activities conducted innocuously and without harm. However, the dual capability of drones to inflict physical damage to property and engage in invasive surveillance raises significant challenges for governments in crafting legislation and regulatory frameworks governing UAV licensure and permissible activities (Horsman, 2016). Additionally, the potential for drones to be utilised for illicit purposes, such as smuggling contraband into prisons without detection by security personnel, underscores the extent to which UAV technology facilitates criminal activities (Yaacoub et al., 2020).

As drone technology continues to advance, with enhancements in flight range, payload capacity, and endurance, the prevalence of such applications is expected to rise. However, concomitant with these advancements is the increased exploitation of drones by criminal elements for nefarious activities. While China leads globally as the foremost exporter of armed drones, the widespread availability of commercial UAV technology has prompted numerous countries to recognise the strategic benefits of integrating drones into their military operations (Marcus, 2022).

2.2 Types of UAVs

Towards the latter years of World War II, the evolution of aircraft towards more sophisticated models prompted the implementation of regulations governing UAVs. Since then, UAV technology has evolved to encompass drones of various shapes and sizes (Stöcker et al., 2017; Vergouw et al., 2016). The physical characteristics of a drone play a crucial role in aiding law enforcement authorities to discern its intended purpose and ownership. For example, while size serves as a significant distinguishing factor, trained law enforcement personnel can discern between fixed-wing drones and multirotor system drones based on their appearance (Vergouw et al., 2016). The primary distinguishing feature between these types of drones lies in their wing structure, which directly influences their flight characteristics (Vergouw et al., 2016). Moreover, drones vary in weight, ranging from mere grams to kilograms, with size scaling accordingly, from imperceptible to the naked eye at a distance to the size of a full-fledged commercial aircraft (Vergouw et al., 2016).

The MQ-1 Predator UAV emerged as one of the pioneering weaponised drones employed in warfare. Initially utilised for surveillance and reconnaissance purposes, its role evolved with the integration of military missiles (Marcus, 2022). Its successor, the Hunter-Killer MQ-9 Reaper, was specifically designed to fulfil its namesake function. With dimensions measuring 12 feet in height and 36 feet in length, the Reaper features a nose-mounted camera, a specialised rear structure enhancing stability, and the capability to deploy laser-guided or GPS missiles. It boasts a maximum recorded speed of 287 miles per hour (Marcus, 2022).

2.3 Terrorist Use of UAVs

Terrorists use of drones consists of a broad spectrum of activities, ranging from surveillance and propaganda dissemination to interception and the disruption of other aircraft (Yaacoub et al., 2020). Extremist groups leverage UAV-led attacks for executing their nefarious agendas by employing tactics such as the dropping of improvised bombs or loitering around critical infrastructure such as airports and military airspace. Notably, the deployment of Vehicle-Borne Improvised Explosive Devices (VBIEDs) poses a significant threat, capable of targeting specific individuals or infrastructure (Yaacoub et al., 2020).

Surveillance constitutes another pivotal tactic employed by terrorist organisations, facilitating the capture of footage to support propaganda campaigns, gather real-time intelligence during attacks, or plan future assaults. The Ministry of Defence highlights Russia's possession of multifunctional, long-range drones capable of participating in coordinated swarms and executing precision airstrikes, with the integration of AI amplifying their destructive potential (Bendett, 2021).

Russian military activities have focused on the development and deployment of drone swarms across various domains, including air, ground, and maritime environments. This strategic investment has bolstered Russia's intelligence capabilities and elevated its prowess in UAV-guided warfare (Bendett, 2021). Drone swarms, characterised by coordinated attacks conducted by multiple drones simultaneously, are increasingly utilised for intelligence gathering, offensive strikes, and defensive operations (Marcus, 2022).

The convergence of drone technology with advancements in AI has propelled the growth of drone swarming, with applications ranging from intelligence gathering to offensive and defensive missions (Abdyraeva, 2020). Armed forces worldwide are striving to enhance autonomy for UAVs, presenting both opportunities and threats in future conflicts (Abdyraeva, 2020). As stated, Russia, in particular, has spearheaded the development and integration of drone swarming into its military arsenal, signalling a shift towards the utilisation of robotic systems over conventional human forces and fundamentally altering the landscape of modern warfare (Abdyraeva, 2020).

3 Challenges Related to the Malicious Use of Drones

3.1 Forensic Challenges

Detecting the malicious use of drones poses a myriad of forensic challenges, particularly concerning the digital investigation process (Horsman, 2016). One of the primary allurements for purchasers of UAVs is the inclusion of camera features, significantly expanding the scope of activities that can be undertaken with the drone. In investigative contexts, gathering camera footage assumes paramount importance in identifying and analysing the purpose of travel and the trajectory of the flight path (Horsman, 2016). The data embedded within images or videos yields crucial geo-location information such as that related to the drone's origin and intended destination. Similarly, analysing flight data is instrumental in discerning the designated flight path of the drone (Horsman, 2016).

However, extracting digital data from drones presents inherent challenges, compounded by issues of maintaining evidential integrity. As a result, this poses a formidable task for digital forensic investigators and LEAs. Moreover, discrepancies in data recording methodologies among drone manufacturers further complicate the investigation process, potentially rendering flight data inaccessible (Horsman, 2016). In certain instances, data acquisition might be impeded altogether due to connectivity issues with the device. Even in cases where access to the device is feasible, conducting live data extraction might compromise the integrity and admissibility of the evidence (Horsman, 2016).

Despite advancements in forensic recovery and investigation processes, UAVs harbour various technical and operational vulnerabilities that render them susceptible to exploitation. Connectivity and communication play a pivotal role in the functionality of drones, with wireless technology integrating them into the Internet of Things (IoT) ecosystem. While this facilitates seamless connectivity akin to smartphones connecting to Wi-Fi networks, it also exposes drones to vulnerabilities that can be exploited by hackers (Guitton, 2021).

UAVs heavily rely on geolocation networks for navigation, rendering them vulnerable to attacks targeting GPS signals aimed at hijacking the unmanned vehicles

(Guitton, 2021). Wi-Fi jamming stands as a prevalent method of intercepting drones, where the connection between the drone and the controlling device is compromised, leading to hijacking (Yaacoub et al., 2020). Moreover, drones are susceptible to malware infection, potentially altering communication between the UAV and its controller. The unencrypted, wireless transmission of monitoring data further exacerbates vulnerabilities, increasing the risk of interception (Yaacoub et al., 2020). Spoofing poses another significant threat to UAV technology. It involves the collection and modification of data transmitted through the drone feed, allowing hackers to gain control of the vehicle (Yaacoub et al., 2020). Such interceptions can result in altered flight paths, manipulation of travel objectives, or even malicious actions orchestrated by criminal actors (Yaacoub et al., 2020).

The size of certain drones poses a significant challenge for detection on flight radars, particularly when wireless and satellite communications are disabled. In such cases, visual identification becomes the sole means of detection, but the task is compounded by the proliferation of UAVs with diverse characteristics. Moreover, the dynamic nature of drone movements further complicates visual identification, hindering efforts to attribute responsibility in the event of an attack (Guitton, 2021). Similarly, navigating the response to destructive drone incidents presents a formidable challenge for law enforcement and military personnel, primarily due to the autonomous nature of UAVs. Autonomy in this context refers to drones operating without direct human control, raising accountability concerns regarding the allocation of consequences (Horsman, 2016).

Autonomy can be achieved through AI or pre-programmed flight paths, with militaries worldwide attempting to enhance autonomy within drone swarms. This advancement enables UAVs to collaborate and make real-time decisions independently, without human intervention (Abdyraeva, 2020). As a result, the burgeoning autonomy in drone technology is a focal point in current research and literature, with the increasing use of fully autonomous drones and drone swarms poised to shape the future landscape of drone usage and warfare (Lehto & Hutchinson, 2020).

3.2 Privacy and Security Concerns

Over the course of three years, the number of incidents involving encounters between UAVs and airplanes surged from 6 to 93 occurrences (Yaacoub et al., 2020). This sharp increase has spurred governments and researchers to confront the inherent dangers to privacy and security posed by drones, highlighting the inadequacy of current approaches to addressing UAV misuse (Yaacoub et al., 2020). Therefore, this escalating malicious use of drones for both physical and cyber-attacks not only raises safety concerns but also threatens national security and compromises the privacy of both targets and victims (Yaacoub et al., 2020).

Key safety concerns revolve around the absence of secure features on UAVs, leaving them susceptible to loss of control or inability to shut down the device.

Operational features also pose risks, with malfunctions potentially resulting in collisions with other airborne objects and causing potentially catastrophic outcomes (Yaacoub et al., 2020). These safety vulnerabilities, stemming from potential errors, also serve as potential avenues for malicious exploitation of drones. For instance, weaknesses in technical and operational performance render UAVs vulnerable to hijacking (Yaacoub et al., 2020).

The challenges regarding privacy stem from the potential for drones equipped with cameras to trespass over private property, encroaching upon individuals' personal freedoms. In certain instances, the aerial surveillance conducted by drones over residential or private areas might violate privacy laws, as the collected videos and images can be exploited for blackmail or fraudulent purposes (Yaacoub et al., 2020). Furthermore, commercial applications of drones for surveillance and tracking individuals raise concerns about privacy breaches, particularly when individuals are unaware of being monitored or targeted for tracking, essentially amounting to stalking (Yaacoub et al., 2020).

In terms of security, drones are frequently utilised to target and intercept Wi-Fi and other wireless devices. The assumption that drones cannot reach these devices often leaves them unprotected by legal safeguards (Yaacoub et al., 2020). Consequently, reliance on single-factor authentication and the use of common passwords render wireless networks vulnerable to exploitation by drones, providing unauthorised access to these networks (Yaacoub et al., 2020). The capability of drones to approach buildings closely in order to gain access to different networks heightens their appeal to criminals. Moreover, the susceptibility of drones to hijacking by other drones underscores the weakness of authentication systems, thereby allowing malicious actors to compromise control over drones easily (Yaacoub et al., 2020).

4 Current Laws and Legislation Overseeing the Use of UAVs

Current guidelines and regulations governing UAVs vary across different jurisdictions and internationally. Previous regulatory procedures have outlined four primary goals for the approval and notification process (Stöcker et al., 2017):

- Andcknowledgment of local or regional operational limitations.
- Prevention of disputes between different airspace users.
- Reduction of public concerns and disruptions.
- Enablement of effective monitoring of planned UAV flights

The implementation of these goals aims to establish regulations that cater to the interests of all stakeholders. The principal regulatory authority in the aviation industry is the CAA, as per the regulations outlined in 2015. Notable articles within these regulations include (Horsman, 2016):

> *Article 167(2)*, which imposes restrictions on how close UAVs can approach certain airspaces or infrastructure. Examples include maintaining a distance of at least 50 meters from individuals, vehicles, or structures, and a prohibition on flying over or within 150 meters of densely populated areas.
>
> *Article 138*, which addresses the restriction of pilots from knowingly or deliberately endangering people or property during flight.
>
> *Article 166(3)*, which mandates that UAVs must maintain a line of visual sight during flight to ensure traceability and prevent collisions. (Horsman, 2016)

In the UK, Robert Knowles became the first individual to be charged for flying a UAV dangerously and illegally, resulting in a fine of £800 for breaching the regulation of maintaining a 50-m distance from a structure (Horsman, 2016). This case underscores the legal complexities surrounding UAV usage, as it intersects with various pieces of legislation. For instance, incidents involving the destruction of property might invoke criminal law, such as the Criminal Damage Act 1971. Moreover, instances of privacy breaches could potentially fall under the ambit of laws such as the Protection from Harassment Act 1997 and the Sexual Offences Act 2003 (Horsman, 2016).

However, the challenge extends beyond national borders, as regulations governing UAVs vary transnationally. While some states might prohibit their use, others may adopt more lenient approaches (Custers, 2016). The affordability and accessibility of drones in regions with lax regulations exacerbate the difficulty of preventing the proliferation of illegal technology in open societies (Custers, 2016). Furthermore, the issue extends beyond mere flying regulations and ownership guidelines; it pertains to the lack of enforcement concerning data and human rights legislation, compounded by a dearth of global discourse on the matter (Lee et al., 2022). Therefore, addressing these challenges necessitates comprehensive regulatory frameworks and international cooperation to mitigate the potential risks associated with UAV usage.

5 Recommendations

The widespread adoption of Unmanned Aerial Vehicles (UAVs) for both recreational and commercial purposes presents a novel and intricate challenge for policymakers and governments as they grapple with addressing the myriad challenges drones pose to national security and society. While privacy and security concerns are paramount, recommendations for addressing these issues will hinge on whether to bolster the protective features and technical capabilities of UAVs or to tighten regulatory restrictions.

Recent trends indicate an increasing reliance on digital forensic methods in UAV investigations. However, these investigations often lack sophisticated tools to detect and collect evidence effectively, highlighting the need for specialized law enforcement units trained in handling UAV attacks and misuse (Guitton, 2021; Yaacoub et al., 2020). Guitton emphasizes the necessity of developing specific legal frameworks tailored to the unique challenges posed by drones, given the exponential growth in drone manufacturing and the subsequent decrease in production costs (Guitton, 2021).

In light of these considerations, there is merit in adopting a more systematic approach to UAV-related investigations. Rather than relying solely on existing investigation procedures, there is a call for the development of a new framework dedicated to addressing UAV attacks and devising appropriate response strategies. Such a framework would ensure that law enforcement agencies are equipped with the requisite expertise and resources to handle UAV-related incidents efficiently (Guitton, 2021). Ultimately, the formulation of normative and legal frameworks tailored to the evolving landscape of UAV technology is imperative to safeguarding national security and protecting societal interests.

Drones, essentially aircraft, pose a significant threat when utilised maliciously, with their targets often being physical infrastructure vulnerable to aerial attacks (Guitton, 2021). Therefore, to enhance security measures, particularly in critical environments such as airports and military installations, increased surveillance and monitoring of online purchase histories and import patterns are essential for early detection and prevention of UAV threats. Similarly, strengthening laws and regulations governing drone licensing could also aid in tracking autonomous UAVs. However, the challenge remains that criminal elements might still exploit unlicensed drones for nefarious activities (Yaacoub et al., 2020).

In response to these challenges, innovative approaches are emerging, such as the UK government-supported cybersecurity boot camp, where students are trained in defending against cyber-attacks and identifying vulnerabilities in drone technology (Yaacoub et al., 2020). This initiative not only provides insights into current vulnerabilities but also informs necessary improvements in drone security. Furthermore, continued technological experimentation and collaboration between public and private sectors are imperative to address these evolving threats. In addition, effective policies must be crafted to guide technological advancements in the context of modern warfare (Bendett, 2021). By mobilising resources and expertise, stakeholders can work towards safeguarding national security in the face of drone-related threats.

Intrusion Detection Systems (IDS) have emerged as a recent method for monitoring and detecting the malicious use of drones (Yaacoub et al., 2020). These systems analyse network data to identify anomalies that might indicate cyber-attacks, particularly threats such as malware and spoofing, with the aim of preventing system compromise (Yaacoub et al., 2020).

One approach within the UAV environment is the implementation of rule-based intrusion detection schemes to safeguard communication between the UAV and the base network. While studies have shown rapid detection of attackers within 40s, rule-based systems face challenges in detecting previously unknown attacks, as they rely on predefined rules and might miss unfamiliar anomalies (Yaacoub et al., 2020). Anomaly-based detection systems offer another avenue for preventing attacks such as jamming, Denial of Service (DoS), and drone overheating. Initial results indicate accurate detection of anomalies related to DoS attacks even though further testing is required to validate the effectiveness of this method (Yaacoub et al., 2020).

Overall, IDS represent a promising tool in the defence against malicious drone activities, providing real-time monitoring and detection capabilities. Continued research and development will be crucial in refining these systems to effectively mitigate emerging threats in UAV technology.

6 Conclusion and the Future of Drones

The rapid proliferation of drone technology has created a regulatory and response grey area, leaving governments uncertain about how to address emerging challenges. The unpredictable nature of drones has further complicated matters, with governments struggling to offer clear solutions and future strategies. A primary challenge lies in striking a balance between accommodating the needs of UAV users and addressing concerns regarding human rights and privacy (Stöcker et al., 2017). While privacy concerns are paramount, they are often intertwined with security issues, as the personal data collected by drones can pose risks to individual and societal security (Lee et al., 2022). The integration of AI adds another layer of complexity, potentially transforming drones into autonomous units capable of executing tasks independently (Abdyraeva, 2020; Guitton, 2021). While this presents legal and ethical dilemmas, effective government oversight and regulation could leverage AI as a crucial tool for detecting and mitigating malicious drone activities (Guitton, 2021).

Central to developments in this field are advancements in drone swarm technology and the pursuit of autonomy. Drone swarms equipped with AI have the potential to conduct intelligence gathering and execute impactful attacks. Therefore, as these technologies continue to evolve and become more sophisticated, their significance in modern warfare cannot be overstated (Vergouw et al., 2016).

In the years ahead, various stakeholders with divergent motivations will vie to shape the trajectory of drone usage. However, for governments to effectively mitigate the risk of malicious drone activities, they must develop a clear vision for the future of UAVs. While existing legislation, such as regulations set forth by the CAA, provides a solid framework, both national and international aviation authorities are recognising the need to update regulations to align with evolving customer needs and technological advancements (Stöcker et al., 2017). Traditionally, UAV regulations have been reactive responses to incidents involving malicious drone use. However, with advancements in knowledge and understanding of this domain, policymakers have an opportunity to take a proactive approach (Stöcker et al., 2017). Nonetheless, it is evident that the pace of technological development in the drone industry outstrips that of legislation and regulations. Therefore, it is imperative to adopt a flexible stance that prioritises not only national security but also the safety and privacy of civilians (Stöcker et al., 2017).

References

Abdyraeva, C. (2020). *Drone Swarms—A Future Threat to Armed Forces?* European Army Interoperability Centre. https://finabel.org/drone-swarms-a-future-threat-to-armed-forces/

Bendett, S. (2021). *Strength in Numbers: Russia and the Future of Drone Swarms*. The Modern War Institute (MWI) at West Point. https://mwi.usma.edu/strength-in-numbers-russia-and-the-future-of-drone-swarms/

Custers, B. (2016). Flying to New Destinations: The Future of Drones. In *The Future of Drone Use: Opportunities and Threats from Ethical and Legal Perspectives* (pp. 371–386).

Guitton, M. J. (2021). Fighting the Locusts: Implementing Military Countermeasures Against Drones and Drone Swarms. *Scandinavian Journal of Military Studies, 4*(1), 26–36.

Horsman, G. (2016). Unmanned Aerial Vehicles: A Preliminary Analysis of Forensic Challenges. *Digital Investigation, 16*, 1–11.

Lee, D., Hess, D. J., & Heldeweg, M. A. (2022). Safety and privacy regulations for unmanned aerial vehicles: A multiple comparative analysis. *Technology in Society, 71*, 102079.

Lehto, M., & Hutchinson, B. (2020). Mini-Drones Swarms and Their Potential in Conflict Situations. In *15th International Conference on Cyber Warfare and Security* (Vol. 12, pp. 326–334).

Marcus, J. (2022). Combat Drones: We Are in a New Era of Warfare—Here's Why. *BBC News*. https://www.bbc.co.uk/news/world-60047328

Mitchell, R., & Chen, I. (2014). Adaptive Intrusion Detection of Malicious Unmanned Air Vehicles Using Behaviour Rule Specifications. *IEEE Transactions on Systems, Man, and Cybernetics: Systems, 44*(5), 593–604.

Stöcker, C., Bennett, R., Nex, F., Gerke, M., & Zevenbergen, J. (2017). *Review of the Current State of UAV Regulations. Remote Sensing* (Vol. 9, No. 5, p. 459). MDPI AG. https://doi.org/10.3390/rs9050459

Vergouw, B., Nagel, H., Bondt, G., & Custers, B. (2016). Drone Technology: Types, Payloads, Applications, Frequency Spectrum Issues and Future Developments. In *The Future of Drone Use: Opportunities and Threats from Ethical and Legal Perspectives* (pp. 21–45).

Yaacoub, J., Noura, H., Salman, O., & Chebab, A. (2020). Security Analysis of Drones Systems: Attacks, Limitations, and Recommendations. *Internet of Things, 11*, 100218.

Printed in the USA
CPSIA information can be obtained
at www.ICGtesting.com
CBHW051917180824
13381CB00003B/82

9 783031 622274